Telescope

The History of the World's Most Famous Telescope

(Everything You Need to Know for Astrophotography or Visual Use)

William West

Published By **Jessy Lindsay**

William West

Telescope: The History of the World's Most Famous Telescope (Everything You Need to Know for Astrophotography or Visual Use)

ISBN 978-1-77485-449-5

Legal & Disclaimer

The information contained in this book is not designed to replace or take the place of any form of medicine or professional medical advice. The information in this book has been provided for educational and entertainment purposes only.

The information contained in this book has been compiled from sources deemed reliable, and it is accurate to the best of the Author's knowledge; however, the Author cannot guarantee its accuracy and validity and cannot be held liable for any errors or omissions. Changes are periodically made to this book. You must consult your doctor or get professional medical advice before using any of the suggested remedies, techniques, or information in this book.

TABLE OF CONTENTS

Introduction

Astronomy is an amazing and enjoyable hobby, with so many amazing things to see that many lifetimes will not be enough time to see and discover all the stunning objects that await the eyes of the budding amateur astronomer.

It's only natural to desire to purchase an eyepiece as soon as possible and the bigger the better. This is often where logic and common sense meet.

It's easy to be caught up in advertisements in glossy magazines that make various claims. You could ultimately purchase something that is nothing more than a toy , or spending too much cash on a scope which isn't suited to your requirements or rarely used.

I'm hoping that the info included in this book is helpful in helping you select the right telescope which can be useful to you for many years to come and will not break the bank.

Chapter 1: What If I Really Do Not Need A Telescope?

Okay, we've come to the point that you can't live without a telescope. However, before you make the leap there are plenty of things to think about.

What is the maximum amount I can spend or am I permitted to spend?

What kind of telescope do I require (this is based on your preferences)?

What kind of stand or mount would most suit my needs?

It will be employed for star gazing on occasion or on a more frequent basis?

What is it that the telescope designed to be employed?

Is it going to be permanently installed perhaps in a garden or outside, and will the structure be taken inside at the end of every session?

Do you want the telescope to be transportable or portable?

What is the maximum amount I can spend, or am I permitted to spend?

I have no apologies for asking this question twice. it's amazing how expensive telescopes can become just expensive decorations and never get to be able to see the stars they were made for.
How do you decide? Every person is unique, so the best option isn't always available and the majority of options are a result of compromise of one kind or another , typically financially.
For occasional stargazing , a good option is the spotting scope utilized by birdwatchers. There are some that are superb. They offer a wide and bright field of view and usually feature apertures of more than 60mm. Most have zoom eyepieces.

These are fantastic to get general perspectives of star clusters, as well as for taking a look at of the Milky Way. The Moon is also stunning when observed through the telescope for spotting. The views of the planets aren't great, however the planets' inner regions will be visible as discs (or an elongated crescent when it comes to Venus as well as mercury) Jupiter and Saturn show clearly as discs along with the Galilean Satellites Jupiter are readily visible as is Saturn's biggest moon Titan.

A spotting scope is easy to be mounted on a tripod for photographers and it can be carried over your shoulders and carry it wherever you go. It can be used for many purposes other than astronomy and bird watching. The addition of a mirror or prism in the focal point redirects the light towards an easier viewing place.

Different types of telescopes.

For the typical astronomer, there are three widely available types of telescopes. A browse through magazines like Sky and Telescope or Astronomy magazine will give you a number of instances of the three fundamental kinds.

A refractor that uses lens

A reflector that makes use of mirrors

A compound telescope that makes use of both

Like everything else, there are many technical terms and phrases that are used to describe telescopes and their capabilities. They are the ones you need to know before buying the telescope. They are fairly easy to understand, but it's important to know the significance of each to be able to make informed choices regarding the purchase of a costly item.

Some terms used in telescopes and what they are:

Diameter

It is usually expressed in centimetres now and refers to the dimension of the lens's primary, mirror or corrector plate thus an old-fashioned 8" Reflector can be transformed into 20 centimetre or 200 millimetre reflector.

Focal ratio

There are many terms focal ratio in the description of the telescope of any kind. The focal ratio is the size of the mirror or objective lens multiplied by the length focal of the lens or mirror or , in other words, the distance between the mirror or lens where the image is taken to a focal point.

It is shown in the form F/D, for an example, a lens with 1500mm of focal length and 100mm diameter would have F15 as its focal ratio however a focal distance of 1000mm and 100mm in diameter will give the focal ratio F10.

The lower the focal ratio, greater the area of view. Thus, the field of vision of a telescope that has F10 as a focal ratio is larger than that of one with a focal rate of F15.

Magnification

Do not be enticed by absurd claims of magnification. While theoretically, the claims of magnifications may be achievable, you might not be able to utilize these. It will all depend on your primary optics, the performance of the eyepieces, your sight (which refers to the quality and the conditions for observing particularly the turbulence in the atmospheric conditions during an astronomical view) as well as the quality of your telescope mount and your place of residence.

Experience and perception what you can see will determine the best magnification for any object. If the image gets slightly blurred, it's best to lower the magnification. The magnification of an optical device is measured by the length of its object multiplied by its focal length of the eyepiece. For instance, the case of an objective lens mirror has the focal length 1500mm, and its focal distance of lens is 25mm, then the 1500/25 is 60 or 60 X60. Therefore, eyepieces can offer different magnifications based on diverse focal ratios. Of course , there are numerous types of eyepieces . We will discuss this in more detail in the future.

Optical quality

The precision with the point at which a lens or mirror is made will have the biggest impact on its ability to give a clear and sharp image. This is commonly referred to as Rayleigh's limit . If we recognize that the frequency of light yellow is approximately 580nm (nano metres (or 5800 Angstrom) in order to get acceptable optics, there must be no deviation from the curve that is greater than 580nm/2 will be permissible for a lens or 580nm/8 in the case of a mirror. which is known as 1/4-wave optics.

You can clearly see, the requirements are more stringent for lenses than for mirrors and, luckily, this information will be in the specifications of telescopes. You'll often see mirrors that are described as 1/8 wave and even 1/10. They will be at a cost that is higher because the work involved to create this kind of precision is lengthy and costly.

Magnifications and focal ratios that are large are typically associated with refracting telescopes, so let's take a look at the refractor first.

Refractor.

A refracting scope is the image that pops up whenever the term telescope comes up. The refracting telescope is one of the oldest telescope type, first created by Jan Lippershey. It was later became popularized by Galileo who enhanced his original concept and applied the design to study astronomy.

The refractor is comprised of the eyepiece and lens The lens is the primary one. It collects and is able to refract (bends) sunlight into a point that is then increased by an eyepiece.

The main issue with refractors is chromatic distortion that can be eliminated to a significant extent by creating the front lens of two glass pieces (a doublet). It is composed of two different types of glasses (crown and flint). Both have different refractive indexes, meaning that they reflect light with different quantities) specifically chosen to minimize the number of false colours in the image. Since a single lens can never be able to bring all wavelengths light into the same focal point. Because of the sheer number of optical surfaces that need to be made, refractors are expensive devices.

There are a lot of small refractors available for sale in markets, however, typically they're not of good quality since they are often manufactured in bulk to lower the price . Most people who purchase them never explore their astronomy interests further. A quality refractor is expensive and any object with a diameter less than 75mm should be considered an item toy. Most commonly, the focus ratios range from F12 to F15.. Anything lower than F12 suffers from extreme color aberration (false color).

Refractors that are more expensive size for dimensions come with a doublet constructed from Flourite EDL (extra lower dispersion) glass, and premium Schott glass. It is no surprise that they are priced at a higher.

There are some fantastic short-focus refractors with F8-10 focal range and generally 100 mm in diameter. However, the main lens is composed of fluorite, and they aren't cheap for a telescope that is just starting out.

Refractors provide excellent moon views, however when you are observing the sun, the safety measures of projecting the image onto a card or using a special filter, or solar observation device purchased from an astronomical retailer that is reputable must be foremost in your mind. Do not utilize the solar filter that comes as an extra included with smaller refractors. They are extremely dangerous since they could break very easily and cause blindness for the user.

Refractors are great for planet observations since they offer high magnifications because of their extended focal length.

The Newtonian Reflector

The reflector was designed in the hands of Isaac Newton the reflector is the most well-known model of amateur instrument. it is the most popular type of amateur instrument. Reflector offers many features that make it superior to the refractor and aperture for aperture it is significantly less expensive. Reflectors have lower focal ratios, and therefore have larger areas of view.

The term "reflector" suggests, reflect the light it receives to bring it to a point of focus. The path of light then diverted through a flat mirror, where it is then brought to sharp focus and magnified with an eyepiece.

The reflector typically has focal ratios of F5 to F8 However, below F5 suffers from spherical aberration. This is because the light rays of all wavelengths do not align to the same spot. This is due to the nature of the mirror's curve isn't spherical, but actually is a parabola. The smaller the focal length, the more deep the parabola has to be and , consequently, more expensive to make since this is a precision job.

Reflectors with a long focal length (F8 and over) are not affected as much , however the telescope can begin to get heavy and difficult to use due with the size of the tube. A huge mount and huge amount of space is an important factor. If you are planning to move the scope around, the bigger it is, the more issues it poses to the user and many reasons to not utilize it. Reflectors with a refractor of F8 and higher are more likely to be utilized for planet observation.

Coma (also known as comet tail)

This happens when the individual light rays of light aren't being reflected back to the same location. If you look in a different direction from the field of vision, the object appears to be wedge-shaped. The effect is more severe when it is further away from the central plane in the scope.

Schmidt-Cassegrain

The Schmidt-Cassegrain telescope is extremely popular over the last 40 years, especially within the United States. It offers numerous advantages of the refractor as well as the reflector. A large corrector plate is used to replace the lens on the top of light paths (the correction plate is curved with a gentle curvature) and also a massive primary mirror.

The light is reflected back through this tube until it reaches a mirror that is mounted to the plate for correctors that reflect the light back into the hole in the main mirror, where it is brought to point of focus and magnified with the eyepiece.

In humid climates, it is necessary to have a dew shield to avoid condensation on the huge corrector plate that is located at the front.

The Schmidt-Cassegrain is a great alternative over both refractors and the reflector. It permits a high focal-ratio usually F10 inside a compact tube since it folds the path of light.

The Schmidt-Cassegrain is extremely portable (to an extent) when compared to a larger refractor or reflector. It also comes with many different accessories that are available to improve the performance and efficiency of the.

It is enough to peruse those pages in Sky & Telescope, Astronomy or any other astronomy journal that is available on newsstands, to witness the popularity of the telescope and the variety of instruments available for purchase.

The most famous models of Schmidt-Cassegrain telescopes have been made through Celestron as well as Meade.

There are various kinds of reflectors like the real Cassegrain as well as the Maksutov. These are however highly specialist instruments and are thus beyond the scope of the present research.

Telescope mounts

Equally important to an optically good telescope is the mounting. An unsuitable mounting can derail the quality of observing with telescopes sporting top optics. The reverse is not the case.

There are two fundamental kinds of mounting: the altazimuth mount and mount that is equatorial. There are a variety of different versions of both. Let's take a look at both.

The Altazimuth Mount

The most simple type of mount available and is the one that is most commonly used with telescopes that are cheap.

The system is based on the elevation of the object over the observers horizon as well as

its Azimuth, the angles of an object in relation to the horizon, measured from north, and then increasing to the west.
The altazimuth mount is always going to have to be adjusted in two planes. The azimuth and the altitude (up and down, and from side to side). It can be very tedious watching, so you can purchase an altazimuth-specific mount that is driven by computing on both axes's. However, it will increase the cost of the telescope. It could be worth looking into an Equatorial mount. A lot of compound telescopes can be converted for Alt/Az to use with the use of an wedge that is equatorial (see an article on fork mounting).
The altazimuth mount is basic and restricted, for example, you will not be able take photos of long exposure, even with computer tracking because of field rotation.
Altazimuth mounts can be useful for those who only need moderate to low power for star clusters and galaxies. By using an altazimuth mount,, one can enjoy a relaxed observing place.
Modernized version of the altazimuth mount referred to as Dobsonian mount.

Dobsonian mount is extremely popular to mount Newtonian reflectors. The mount itself is easy to build and allows a the use of a large telescope placed in a small space. Even big telescopes can be utilized effectively in this setup.

The mount that is equatorial

The equatorial mounting system is an improved method of tracking objects using the aid of a telescope. The mount is aligned with that of the North Pole (or South Pole) of the sky. This means that you only need to follow only one direction.

Sky appears to revolve around the poles. If the axis of one is aligned to that of the North or South Pole of the sky the telescope will follow the axis. This is also known as Right Ascension or RA and is measured in hours minutes, Seconds and Minutes from an unmoved position. The beauty of this system is the fact that the skies appears to complete one complete revolution within the course of a period of time (almost).

An additional axis of 90 degrees to the Polar axis allows that the telescope be tilted North to South which is known by the term Declination (also known as Dec. Declination is measured from the celestial equator , which represents the projected area of earth's actual equator on the celestial sphere. It is measured from the zero degree (on the Equator) from +90o at North celestial pole and 90o at the south celestial pole.

To give an example, to illustrate, the star bright Vega is located with the following coordinates: RA: 18h36m 56.3s and Declination +38deg 47'01". If you are aware of that the Ra or Dec for even the most tiny object, you can locate it easily through using the setting circle method. But despite being included with telescopes along with a sheets of instruction on how to use them, the majority of them aren't large enough for them to have much use as well. The polar axis must to be aligned precisely to the pole in order for this to work correctly.

Accurate Polar alignment can be achieved through practicing, but an easy method to get around this age with "computer everything" is the multitude of computer-based trackers (such like that of the GoTo system) that are readily available for even small-sized telescopes. They have the complete polar alignment software which locks on to a number of prominent star clusters in the database and carries out all the maths for you.

The computerised systems have hundreds, if perhaps thousands of objects stored in their database so you simply have to pick the object you would like to observe then the telescope align itself towards that object. You can input coordinates for objects that are not listed found in the database like minor planets, novae and add other objects you might like to see once more.

The inclusion of a computer system like this is another thing to think about when deciding the amount you want to invest.

Mountings

Two of the most popular kinds of equatorial mounts are is the German mount as well as fork mount. Fork mount.

The German mountain

It is the German mounts are the most popular kind of Newtonian refractors and reflectors. The telescope is mounted on one side of a shaft. It is balanced with counterweights at the back. The German mount must be carefully balanced using the use of a German mount. The eyepiece could become very difficult to maneuver into places, and the inclusion of the "slip band" allows the tube to rotate within its cradle, without the necessity of having it balanced. The combination to both motors makes tracking simple as well. The German mounted is the most very popular option for amateurs.

The fork mount

This mount has become a favorite among Schmidt-Cassegrain as well as other scopes with compound optics. The major benefit of this mount is that it's extremely small. The commercial manufacturers of this kind of instrument include in the mounting the complete motor drive system, making it easy to follow the stars, take photos and utilize all kinds of accessories effortlessly.

Eyepieces

A number of poor eyepieces have damaged a quality telescope! When you purchase an instrument, you'll usually receive in the box the eyepieces. This can be good or bad it is all dependent in the grade of lenses that you receive. For long focal length telescopes like refractors, a less complicated eyepiece can be utilized but in the end, you get what you get for your money.

Small eyepieces made of barrels like 24,5mm won't perform to an acceptable standard, so 32mm (1 1/4 ") diameter is the minimum you need and this is the diameter which is typically offered by the best telescope makers. If you own a big telescope, then the 50mm (2") eyes could be beneficial, however they are priced at a higher, which is a significant higher.

Understanding the basic properties that an eyepiece has is crucial to anyone using the optical instruments, especially when deciding on the best eyepieces to be used on your telescope. The budgetary requirements often dictate the purchase decision when comparing eyepieces and deciding on the best one for your requirements.

There are some terms that we should be conscious of when we talk about eyepieces.

Focal length

We've previously discussed what the lengths of the lens are. the information will be displayed in the barrel or on the side of an eyepiece like we've seen, the same eyepiece can give different magnification depending on the focal lengths.

Field of view

A field of vision is the portion of the field that is observed when looking through the eyepiece. The field of view that is visible through an eyepiece differs dependent on the magnification that is produced by the eyepiece, when it is utilized in conjunction with a mirror or lens (see earlier).

Eyepieces are distinguished from their field stops which is the most narrow aperture through which light that enters the eyepiece has to be able to pass through before reaching the field lens of the eyepiece.

Due to the impact of these variables, the expression "field in view" is almost always either one or the other of these meanings:

Field of view in the actual field

The actual field of sight is size by degrees space of sky that is observed through the eyepiece employed with the telescope (this can differ from instrument to to instrument). The typical figures for this will vary between one tenth one degree to two degrees.

Viewing field of view with transparency

This determines how big the image will appear , and is different from the magnification. This is the same for any fix focal length optical eyepiece, and is a simple way to determine the actual field of view for any particular telescope. The apparent field of view could range between 30 and 110 degrees. The following diagram will simplify understanding of this.

Simulation of different views using telescope eyepieces.

The image on the left shows the image seen through an eyepiece that has a small visible field of vision.

The center image shows the image through an eyepiece with identical focal length with a more apparent field of vision The image is bigger and covers an expanded surface.

The image on the right shows that has the same apparent field of view as the center image, however it has a shorter focal length, which gives higher magnification. This results in exactly the same field

Eye relief

One of the most important aspects to take into consideration when purchasing eyepieces is the eye relief. This is the distance that the eye must be in front of the eyepiece for a comfortable viewing experience. It is dependent on the model of eyepiece, its construction and the magnification. Eyepieces with a long focal length typically offer ample eye relief however shorter focal lengths can cause discomfort when used.

Eye relief.

1 Real image 2 - Field diaphragm Eye relief 4 Exit pupil

An eyepiece that is designed to provide comfortable viewing is recommended to have 5-6mm of eye relief. Anything less is difficult to work with as your eyelashes and the person who is using it must be accommodated. Multi element eyepieces are able to compensate for this , and can even accommodate the wearer of glasses as they might require an eye relief up 20 millimeters.

Huygenian

The most simple eyepiece to be found and is usually comes with a small telescopes that refract. The eyepiece is composed of two convex lenses This design is getting scarcer due to the small eye relief, as well as excessive image distortion and the chromatic aberration. nevertheless very effective for projecting sun-related images onto a card.

The most modern eyepieces are made of multi element design as well as bloomed (coated with a thin layer of anti-reflective coating) to prevent unwanted reflections within the eyepiece from lenses' surfaces, and to increase contrast.

Orthoscopic

This Orthoscopic eyepiece is composed from four parts that provide good color correction, great eye relief, and a small field of view , 40o-50o. It is a general eyepiece. The price of a high-quality Orthoscopic eyepiece is quite low in comparison to more sophisticated designs.

Erfle

The main benefit that comes with the Erfle eyepiece is its large field of view , which is approximately 60 degrees Erfle's feature large eye lenses, excellent eye relief and are user-friendly. They're not the best choice for extreme magnifications due to the fact that they are prone to ghost images, and are also prone to astigmatism.

Plossel

One of the more costly eyepieces, and among the most well-known kinds that is the Plossl makes use of a high quality glass, and comes with at least four perfectly matched elements. The Plossl comes with a 50o field of view, making it ideal for deep sky and planetary observations.

The Plossl is a more compact eye relief than an orthoscopic eyepiece in particular those with shorter focal lengths but it is one of the most sought-after kinds of eyepieces in the market for amateurs.

There are a variety of other kinds of eyepieces available offered to amateur astronomers, including Ramsdens, Kellner, Monocentric, Konig, RKE, and Nagler. It's true that eyepieces, their designs and function could create the capacity to be a whole lot more than them.

Barlow lens

While it is not an eyepiece, it is typically included with telescopes, and is an important part of the toolkit of astronomers.

It is also known as the Barlow lens is a negative lens which is installed between the eyepiece as well as the telescope. It is able to double or triple magnification. If you have only two eyepieces, you can purchase an Barlow lens and come with four.

The benefit is that with the use of a Barlow lens, you can get the highest magnification by using an eyepiece with a longer focal length that is simpler to use and offers greater field of vision over the standard single eyepiece with a shorter focal length.

If you purchase all of your eyepieces that are of the same model from a single manufacturer , they are par-focal that means you are able to switch eyepieces with no need to change focus.

The eyepiece is kept in the proper position by a focuser There are numerous kinds, but the most commonly used is the rack and pinion configuration.

Other accessories

Finderscope

One of the most essential accessories is a finder scope. which is a tiny short-focus refactor that has large field of view which allows you to pinpoint an object that you can view through the telescope. The majority of finderscopes that come with

small telescopes are inadequate , and typically 30X10. Find the biggest finderscope can be afforded 10X50 if you can. This will be a huge benefit when trying to find objects in dim light.
Check your finderscope has been aligned as precisely as is possible with the principal axis of the tube. This is easily accomplished in daylight by aligning the telescope with distant objects and then using the adjustment screws to hold the finder. Keep increasing the magnification of your main scope until you have images that are as precise as possible.
Many finderscopes are equipped with gricules or crosshairs, which make the entire procedure a lot easier.
Star Diagonal
A star diagonal is designed for those who own a refractor, or compound telescope. They can be done without an accessory. They are usually offered as a part of the package. They consist of mirrors (or Prism) which is tilted at 45o with respect from the optical axis. it redirects the beams of light to the eyepiece to allow for better seeing,

especially close to the Zenith (directly above the eye).

Mirror diagonals are more affordable to make than prism diagonals and are not prone to mistakes in colour since they reflect light. They can also produce the effects of mirror reverse, which is when images are reversed from left to right.

Filters

Filters are utilized to improve the clarity of areas of the moon. The filter screws into the threaded part of the barrel. Filters of different colors can highlight different aspects on the surface of a planet. It's more sophisticated than what we require at the moment, therefore we are able to decide the need to buy a filters until we've gained more understanding.

Solar filters

Every other object try to capture the maximum amount of light we can, but with both the Sun in particular and with the Moon we are blessed with a huge amount of light. My personal view is that the solar filter that come by small telescopes should never be utilized. It is an area where even a tiny telescope can collect more sunlight (and the heat) than it needs. If you're interested in observing the Sun make use of the projection technique or seek advice from a specialist manufacturer of telescopes in regards to the most secure method to accomplish this. They have a selection of Solar accessories that are suitable for use using small telescopes.

Moon filters

They are basically neutral density filters that permit the moon to be seen comfortably by reducing amount of light that is transmitted to the eyes. They are a great addition to an astronomer's kit, particularly near the full moon.

Red torch

White light can ruin your night vision and dark adaptation. Use dim red lights if you require an astronomy chart or make adjustments to your telescope or other equipment.

Computerised trackers and motor drives

As we have previously discussed The main benefit to these devices is it permits the user to continue with astronomy without having to worry about constantly making changes to their telescope. These systems are priced at a higher cost, but with the availability of telescopes from the top manufacturers, they are easily modified to fit the particular telescope.

Photography

One of the most exciting aspects with modern-day telescopes is their ease with the ability to photograph the astronomical objects. While film cameras remain employed, the vast majority of modern astrophotography is taken using CCD cameras. Every astronomy magazine contain images shot with less equipment, which is equivalent to pictures that were taken by professional observatories just few years ago.

The CCD camera, also known as a film camera is connected to the telescope by the focussing mounting mount. The mount is utilized to concentrate the picture onto the CCD chip, or the film plane.

There are many more accessories that telescope manufacturers will be willing to sell, but this is sufficient to start you off in this new pastime.

Chapter 2: First Nights With The Telescope

A telescope purchase is only the initial step. new challenges arise on that first night. What do you need to know about setting up the telescope? Which eyepieces you use for what, and what is the best way to control the computer's controls? What is the procedure to follow prior to observation?

The next pages will lead to help you overcome the challenges that can arise when you first commission the telescope. Check out the information on the telescope you have chosen - whether the alt-azimuth or equatorial, or the computer-controlled, and then skip the other pages.

The Right Place

The telescope is not able to discern through clouds or haze. In theory the telescope can reveal exactly what could be seen with the naked eye, just more clearly.

The selection of the ideal location for observation is, therefore, extremely crucial. Although it's attractive to place your telescope beneath the skylight in your home or through your living room's window but the results will be minimal at most. Much effort was invested in the creation of the telescope to create lenses that are as precise as they can be. With a window pane in contrast optical precision is significance, and therefore large magnifications can't be achieved.

GET OUT!

The simple act of opening the window isn't the solution either: then cold or warm air moves through the window and the image flickers. Even prolonged periods of ventilation don't help: Because the temperature outside changes frequently and it takes a lengthy time to allow the interior temperature to change, particularly in rooms that are insulated. Even on dark surfaces , such as roads or roofs and bridges, warm air rises into the air long after a sun-filled day has finished. This creates a blurred image. You can see this effect by looking with your naked eye during warm days: The air over the street is hot like street lamps at the opposite side of the city appear to be flickering in the evening.

This air turbulence also is the reason for the famous flickering of stars as The distant stars we can only see them as points, which

means that only one light beam is able to reach us. As the air moves through air layers with various temperature or density, it gets constantly deflected as it "dances". The planets are so close to us that not just one , but many rays light can reach us. We can therefore view the disc with the naked eye , as being a serene star, despite the turmoil within the atmosphere. Only when we use the greater magnification of a telescope can we observe that it's covered in dust and has the disc with no specifics.
Because light travels long through the air when you look at objects near the horizon, it is best to be patient to see if your target is as tall as they can be within the skies. The Earth's atmosphere will be able to completely block the view.

It is possible to see an amazing night sky from lights of the city. In downtown, sky appears brighter than the stars thanks to scattered lights from the light sources. Only when you are above the horizon will the stars be seen.
DARK SKY

The most important aspect of air turbulent is light pollution. scattered light from billboards, street lamps or garden lighting that is not directed to its intended destination is not just inefficient energy, and is a source of concern for animals living in darkness as well, but also brightens the Earth's atmosphere i.e. our sky. If the night sky appears brighter and brighter than galaxies and stars behind it, we will never see them regardless of how powerful our telescope might be.

THE PERFECT LOCATION

The ideal site for observation must therefore be at the highest possible elevation to ensure that as little atmospheric as possible is between our eyes and the sky and more in a meadow that is cool than in an asphalted hot parking lot. Also, it should be as far from shining street lights or the streets as far as it is possible. It is no wonder that the closest mountain range , or even the plains that rise up to Namibia are popular locations for amateur astronomers.

The most important thing is the access to the internet. It is possible to observe the moon and planets easily even with lighting pollution. If the air is quiet and good, you can make use of the garden that is behind your house. Galaxies and nebulae require dark night sky. Driving to the edge of town is worthwhile and a tiny telescope will show more than a more powerful one located in the city's central area. If you are able to observe your way to the Milky Way with the naked eye, it's a stunning experience.

STARTGAZING IN THE CITY

Stargazing is also feasible in the city, so it is not close enough to a street light. A dark, shady rooftop terrace or backyard can be enough to view the stars of the Solar System, including our Sun (Safe Solar

Observation) and the planets, as innumerable Double stars and star clusters. Globular star clusters also defend themselves from the effects of light pollution. For massive galaxies and nebulae driving towards a dark place is essential.

A bigger balcony at the top of the floor could be an excellent observation point when you have an unobstructed view of the sky and you are above the level of air turbulent, and only the heat from the home will alter the picture.

KEEP ORDER

An easily accessible observation point offers the benefit that you're likely to utilize it more frequently.

If possible, put everything you require along with the telescope so that you don't need to find them before. Apart from the filters, eyepieces, charts of stars, and batteries for the flashlights and mounts Also, consider a seat along with a drink, and warm clothing - even the summer evenings can be extremely cold. A brief checklist can be helpful to determine whether there are any items missing prior to beginning. Cloudy nights are a great opportunity to make an inventory of the things you'd like to look at ahead of time.

Chapter 3: Set-Up An Alt-Azimuth Mount

An alt-azimuth mounting system works in the same way as the two-way panhead that is found on a photography tripod. In addition to the simple operation, the main benefit of this kind of mount is its speedy and simple setup. However, the tracking of celestial objects can be a bit more complicated, as the telescope has to be moved along two different axes. This is where the setup on the AstroMaster 70AZ has been described.

OE

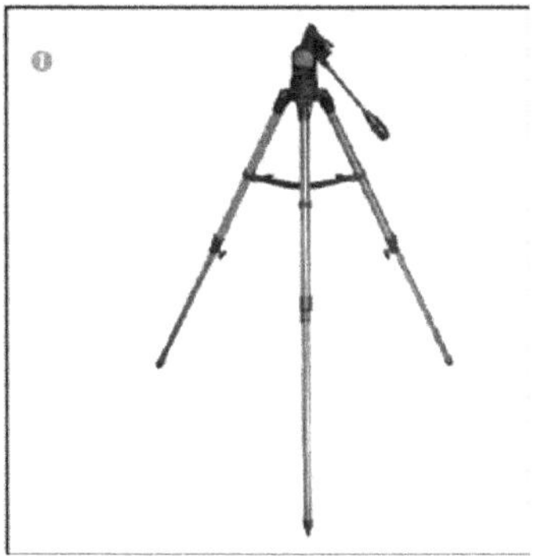

1.) Begin by setting the tripod. It must be level and upright. in a straight line. Lengthen its legs to a height that is comfortable, according to the telescope you are using and whether you wish to view standing or sitting.

2.) Attach the tray for accessories. For AstroMaster, LCM, SLT and SkyProdigy it is secured to the tripod via an rotary motion. With others, it's screwed to the tripod. The plate acts to act as an accessories tray, and can also help stabilize the tripod.
Secure both axes (up/down and left/right) with the screw clamps by tightening them by hand. For AstroMaster telescopes, the long handle functions for clamping screws the height axis. smaller telescopes controlled by computers typically have slipping clutches on the height direction.

3.) Then, place your telescope onto the mounting. generally, it has a dovetail rail that is fixed with two screws through an easy coupling. Particularly with PowerSeeker telescopes the telescope and its clamps are connected to the mount. With the CPC the tube is securely connected with the fork. Certain models come with an adjustable manual for that height angle. The chrome-plated rod should be placed into the holder of the mounting, which acts as a clamp to the object's height (picture near the conclusion of the section). The telescope should be positioned closest to the center of gravity.

4.) Then, remove the caps that protect you and then attach the accessories such as Okulare Zenitspiegel und Sucher - Fertig! including eyepieces, a the star diagonal, and the finder - done!
Make sure to check the finder for a location at the horizon, and adjust it if needed to pinpoint your targets quickly.
For telescopes that have computer control it is now necessary to join the control with your hand (make sure you attach it via the connector marked HC or Hand Control to ensure that you do not cause damage to the electronics as Hand control comes already mounted in NexStar SE). NexStar SE) then then insert batteries or connect to a power supply unit "Tip positive" applies to all present Celestron mounts. i.e. plus is located on the inside of the. If you're using a

power source be sure to not place it or the source of power in the damp grass. Don't turn it off until the hand controller is in place!

If you have a mount controlled by a computer, it is still necessary to complete the alignment, but without having a computer to monitor immediately.
If you own an uncontrolled telescope and you want to use it immediately, place it in the First Objects. If you have a telescope equipped with NexStar Computer Control, you must point it using SkyAlign to three aligned stars to ensure that the telescope is aware of its location under the stars. To make the most from your scope, visit an observation spot that is in the open, which has the lowest light is possible however, it is also easily accessible.

Certain telescopes that are mounted azimuthally come with a fine adjustment of height . This model has the knob for turning is small on the rod is chrome-plated.

Installing an Equatorial Mount

The configuration of the parallactic, equatorial or German mount is a bit more complex than an alt-azimuth mount . After all, the correct ascension axis needs to be set as close with the Earth's orbit. However, tracking the movements of the stars is much easier to do, and you can even utilize the mount for photography in the event that it's accurate enough. The setup is similar to any equatorial mount and can be illustrated with this model: the Omni XLT 150 Newton as an illustration.

1.) Begin by setting the tripod. The pin that is on uppermost part of the tripod should be facing north. It will later be used to make the fine adjustment of the direction of azimuth (right/left). Some of the basic telescopes that are not specifically intended to be used for photography, don't come with a pin since an unreliable orientation to the polar plane is sufficient for visual observation.

2.) Then, attach the tray to the accessory. In certain series the tray is secured to the tripod by the rotary motion or attached to the holders for the Omni series, it's mounted upon the rod that is middle which also serves to secure the tripod to the mount. Remove the azimuth screws from the mount prior to setting it up on your tripod to ensure they don't sit on the pin. attach the tray and mount by using the hand screws that are large. It also spreads and stabilizes the tripod.

3.) Screw in the rod that counterweights and (for the mounts that are not computer controlled) the knobs that turn for both axles. Check the Polar finder (or the hole through which it could be inserted) to see if are able to observe the Polaris. If not, you can adjust the tripod using the screws for azimuth and altitude to be able to be able to see the Polaris star. For this, you will need loosen the mount on the tripod to allow you to turn it around to the side.
The counterweight should be lowered onto the rod that is counterweighted and then clamp it. There is a pin that presses against the rod. The counterweight should be held with the screw that clamps it facing downwards, to ensure that it is able to slide

into the counterweight and you can put it on the rod. Make sure to screw the stop screw back onto the rod at the end of the counterweight rod , and make sure that the axle clamps are tightened hand-tight.

4.) Attach it to the mount. Usually there is no dovetail rail, but it must be inserted into the holder, and then secured by one or two clamping screws. Only for the PowerSeeker models, you have to attach the clamps to the mount first, and then put the telescope in the clamps. In any event, it is easier when you have the tube parallel to ground, and the your clamping screw is pointed towards the upwards. Additionally, you can attach the eyepiece, finder and, if possible the diagonal star. Release the clamp on the right side and rotate the telescope on the axis. Use one hand to hold it. Move counterweight until the telescope is level and stops immediately when the clamp is removed.

5.) Next, fix your R.A. axis and release the declination clamp to align the telescope. To accomplish this, move it around in its

clamps or shift the dovetail bit - depending of the type and model.
Once the telescope is in balance make sure you check the alignment. To observe the telescope it is sufficient to be able to see Polaris at the center of the telescope's polar finder. Then, remove the dust cap off the lens.
For telescopes that have computer control You must now attach the handheld control (make sure to only join it with the connector marked HC or Hand Control so that you don't harm the electronics - Hand control comes pre-mounted in NexStar SE). NexStar SE) as well as then insert batteries or connect to a power supply unit "Tip positive" is the case for all currently operating Celestron mounts. i.e. plus is located on the inside of the. If you're using a power source make sure you do not keep it or the source of power in wet grass. Don't turn it on until your hand controller is in place!
If you own a telescope that is not controlled by computer and you want to use it immediately, place it in the First Objects. If you have an instrument with NexStar

Computer Control, you must point it towards two aligned stars to ensure that the telescope is aware of its location under the stars. To maximize the use from your scope, visit an observation spot out in nature, which is as dark as you can get yet is also accessible.

Chapter 4: Polar Alignment

An Equatorial mount is the most common telescope mount due to the fact that it can only be controlled by only one the axis. This eliminates the issue of field rotation. The camera will always remain in its position with respect to the stars unlike an alt-azimuth mounting, it will always remain in the same position relative with the earth. To work the axis has to be precisely with the Earth's axis.

Simply put, in the northern hemisphere, the mountain is aligned with its pole star (Polaris and Alpha Ursae Minoris), however, not precisely because the star is over half a degree of the pole that is visible to us. For the Southern Hemisphere the alignment of the polar axis is more complex since there is no bright star to serve as an orientation aid near South Pole of the Celestials. Luckily, precise alignment is only required to photograph. If you are looking at visual images with an equatorial mount, it is generally sufficient to align the telescope to the location that correspond to the pole of celestial rays (declination 90deg, right ascension at 0h00) as illustrated in the image below, and then to view through the scope's viewfinder or through the eyepiece. This is a great starting point from which you can begin the tuning. If you are only observing visually and aim your telescope at objects through the viewfinder, instead of using GoTo or coordinates it is likely to be precise enough.

For a rough alignment, it's enough to have the telescope parallel to the Polar the axis (right ascending the axis) and you can clearly see Polaris within the eyepiece. Then, it is possible to follow with the fine tuning of the work.

ALL STAR POLAR ALIGN

The most effective method for aligning polar lines using GoTo-mounts by Celestron is called The All Star Polar Align which will be explained in greater depth in the following. In essence it is a simple alignment using the pre-aligned mount . Then, you move to an erupting star using the software, and all corrections calculated by reference stars are ignored , and the telescope is shifted to the position you would see the stars if it was perfectly aligned. The star then gets centered within the eyepiece by using the alt-az-adjusting screws that are located on the base (not those on the direction buttons on the handheld control). With a higher magnification, or, better yet with a crosshair-type eyepiece you can attain a precise and accurate measurement.

Constellation drawings instead of circles can help you quickly align your mount to the polar plane.

POLAR SCOPE

The Polar scope is by far the most well known method of Polar alignment, however, it's also the most difficult technique to explain. A small telescope positioned on the right ascension axis should be utilized to center Polaris in a position that is slightly off-center. Because Polaris is a celestial pole that orbits as do all stars, the mark should be set to the appropriate date and time. Additionally to this, there are scales for dates and times (Attention Always utilize the standard time, not summer time!) by which the exact location of the clock can be established.

Polar scopes that can be retrofitted onto existing mounts, the scales are typically missing and only some constellations are cut in the scope's eyepiece. In these types of models the right ascension axis that is connected to the polar finder has to be rotated so that the direction to the direction of Big Dipper as well as Cassiopeia when looking with your naked eyes in the skies is similar to the one in the Polar Finder. Be aware that you can't see these stars with the polar finder because the field it covers is smaller than that. This is why you need to examine the sky to the drawing in the Polar Finder. In the event that your alignment appears right then do not alter the axes and then adjust the entire mount until you are at the smaller mark for Polaris.

For the polar scope to be used for use, you need to turn the declination axis to ensure that you are able to see through the polar finder. The declination axis has a hole it, and it is only revealed when it is in the right place.

KOCHAB-METHOD

The Kochab method is similarly to adjusting engravings of the constellation. It relies on the reality that Polaris is Northern celestial pole, and Star Kochab (Beta Ursae Minoris the rear, upper star of the Little Dipper) are on the same line of sight. This is also the case. Polaris must be visible within the telescope of the polar, i.e. the mount has to be facing towards the north.

It is enough to put the Polaris in the center of your circle in the middle so that the center line coincides with the Polaris along with the Kochab. With a bit of practice you can do this in a short time if you gaze at the scope through the polar lens using one eye, and then towards the sky using the other. It is also helpful to rotate the polar axis to the point that a line from the center point is directed to Kochab. This allows you to polar-align the mount using high precision.

DRIFT ALIGNMENT

For permanently mounted mounts utilized for photography, it's beneficial to perform the alignment in accordance with Scheiner. The "drift-alignment" technique is extremely precise, however it is also lengthy and requires a rough alignment.

To accomplish this it is necessary to have an eyepiece that has a crosshair that is aligned with both the R.A. and Dec axes and, when you turn an axis, a sphere in the eyepiece will move through the crosshairs. Then , let the mount run for a few minutes. If the mount isn't properly set up then the star can shift up, down, or in a sideways direction. By moving the mount, you will be able to achieve a more accurate alignment till the object stays exactly in the center for at least 20 minutes. This is not necessary to begin observations, but please take a look at the more detailed explanations on the web.

Full Automatic System: SkyProdigy/StarSense

The SkyProdigy telescopes, introduced in 2011, represent the next phase of the advancement of control of mounts. They come with an integrated camera, through which the controller is able to search for the stars that align with it. The telescope is powered by electricity and an all-round view. Approximately third of your sky must be viewable. Like the most experienced amateur astronomer the mount scans across the sky, and can determine which stars are in the area. Since 2013, it has also the ability to be retrofitted with StarSense-Module StarSense (page 126) for different Celestron telescopes, as of 2016 even for certain Sky-Watcher mounts.

The system known as StarSense is not a GPS This means that you have to input the date, place and time. This information is required for calculating the positions of the planets as well as the horizon. The available SkySync GPS receiver (#93969) will do the calculations on your behalf (although it can't offer anything else). The camera's information is intended strictly for alignment and isn't accessible as an alt-azimuth mount SkyProdigy isn't suitable for photography with astrophotography at all. The hand control is identical to the traditional NexStarHand Control however, it isn't compatible with it since the hardware used to perform the calculations is different than the regular NexStar control.

The SkyProdigy mount is equipped with an integrated camera (left) that can orient itself to the sky. There are a variety of telescopes that can be utilized through its prism clamp.

SkyProdigy telescopes are comparable to SLT mounts and came with comparable telescopes, and different models have entered the market in the past. The setup is similar to the setup of other alt-azimuth telescopes. However, setting up the telescope using the computer control is simpler than those using the NexStar system.

STUDYING with StarSense

In order for use of StarSense technology to be effective it requires an observation point that can let you see at minimum three quarters of the sky. A small backyard isn't enough. Even an astronomer requires more than a handful of stars to be able to determine the direction of his observations. The darker the sky the more luminous. The telescope is rotated clockwise away from its initial position.

After you've set up the SkyProdigy telescope, you must connect the eyepieces, and (except on Newtonian telescopes) the diagonal of the star and the handheld controller, and took off dust covers. it is only necessary to turn to the mount and hit the ALIGN button once it is in place. The telescope will then look for the aligning stars.

SkyProdigy 130 Newton

When you use the telescope first time, make sure to calibrate it in accordance with the instruction manual. This helps eliminate variations in the alignment between the camera and telescope, and also ensures that the objects are aligned in the eyepiece in the future. To achieve this, align a bright star within the eyepiece using the keys for direction and use the StarSense

Camera/Calibrate feature in the menu. After pressing ENTER then the StarSense camera can determine what the telescope's alignment is. It is possible to adjust the StarSense menu also offers an option to alter the camera's sensitivity for different situations, like full moon nights, bright suburban sky as well as dark and cloudy skies. If StarSense is unable to find aligned stars, you can modify the camera's settings.

The other use for the SkyProdigy telescopes is similar to the NexStar hand-control. It is possible to use the Sky Tour key to have it show objects that are visible. When you press UP or DOWN, you can select your destination from the list. As the moment you hit ENTER, the telescope will automatically be slewed to that location. Always start by using the eyepiece that has most focal length because it has the biggest area of vision.

EQUATORIAL MODE

You may also utilize the StarSense module for mounts with equatorial axes however, you need to align the mount with the celestial pole, and then place it on the

marks of the index before starting on the StarSense automatic alignment.
Similar to the standard hand Control You can also utilize the Polar-Align but it is recommended to include a few additional sources of reference prior to. The Polar-Align now approaches another reference star that is the one you need to center in your telescope using the adjusting screws on the mount. Another StarSense Auto-Align will be advised to allow the GoTo-positioning to be accomplished with precision.

The StarSense module includes two viewfinder bases , and replaces the finder on the telescope.

Chapter 5: Accessories: Eyepieces And

Finder

The finder on the telescope can be used to locate objects or aligning stars rapidly. Even the eyepiece that has the lowest magnification will display only a tiny portion of the sky but there are many more stars, which makes the task of locating them difficult. With a high-quality finderscope you can have a vast area of view and quickly identify your target.

Optical finderscopes are a classic solution. They are a small and low-magnification scope with an aiming crosshair to aim at your target or for moving between stars. They're a disadvantage because they aren't able to show the entire sky. Being able to see more stars is useful for telescopes with large apertures or in regions with low star count the sky.

Optic finders are tiny telescopes (left). Finders that use red dots (right) provide an actual target spot between stars with no the need for additional magnification.

A red dot finder is made up of an LED which projects the red luminescence onto glass piece which appears to be floating among the stars. It works even if you view this finder at a certain distance meaning you don't need to place your eye on an eyepiece. Because you are able to observe all the sky in one place, orienting is simple. Along with an eyepiece that offers a wide field of view and lower magnification, it is possible to be able to pinpoint your targets quickly. With the StarFinder Professional, Telrad as well as Rigel Quickfinder have additional options such as more advanced Reticles.

A basic red dot finder with a light point and graduated circles that are found on one of the earlier AstroMaster. It can be adjusted with the help of two screws in silver. The eyepiece of this refractor is positioned in an arc of stars.

ASSESSING THE FINDER

When using the finder, it is necessary to first adjust it. Otherwise, it's likely to not precisely in the same direction as the telescope. Then, point your telescope towards the distance of an object such as an church tower, mountain top, or even a landmark. It is crucial that you are able to view it by your naked eyes. The first step is to select the eyepiece with the longest focal length because it offers the least magnification as well as the widest field of vision. This is the simplest method of locating your target within the telescope. It is then centered using an eyepiece that has a shorter focal length.

Then, look over using the locater: If you notice that the crosshairs are not aligned or the light point isn't on the target, then you need to adjust the finder accordingly. There are three or two adjusters in the finder to adjust the crosshairs. Make sure they are turned until the crosshairs are aligned with the desired target. This allows the finder to be adjusted to allow you to utilize it.

When using the Newtonian the eyepiece holder can be found at the top of the tube. Focusing is accomplished by two wheels made of plastic.

CHOOSING AN EYEPECE

The eyepieces are located in the focuser , or the star diagonal. The majority of eyepieces have the standard size of 1.25" However, there are eyepieces with a 2" diameters that can provide greater fields of view with longer focal lengths. Telescopes equipped with a 2W focuser generally come with an adapter that can accommodate less compact 1.25" lens.

Always begin with the eyepiece having long focal length. Typically, it's in between the 20-32 millimeters. It has the lowest magnification and makes it easier to locate the object. It is important to focus the image by using the focus wheels on the lens;

in the case of Schmidt-Cassegrains the focus knob is situated at the back side of the tube. Every person has to concentrate on their own as nearly everyone has a distinct perception of the world. After replacing an eyepiece it's usually necessary to adjust the focus a bit. Stars are simply tiny points of light inside the telescope. If they are the smallest you've found the focal point.

It is simple to find the current magnification simply by dividing the length of focal lens by the length for the lens:

Magnification = Focal Length(Telescope)/Focal Length(Eyepiece)

For Schmidt-Cassegrains and refractors the stars diagonal (prism or mirror) is provided. It's not only an ideal viewing angle it is also essential for several lens telescopes to get into focus .

The Barlow lens is placed between the eyepiece, focuser or star diagonal, and it will increase the magnification by a specific amount, typically 2x. Start by observing without it and then try it with larger magnifications in the future.

NexStar+ Hand Control NexStar+ Hand Control

The NexStar+ hand control for the Celestron telescopes is a flexible hand control device that gives the "intelligence" to the mount. It is not just used for alignment however, it also manages the telescope and is able to adjust the telescope to make the mount function as smoothly as is possible. From the language settings of the menus as well as your own catalog of objects to GoTo-speed compensation and gear play You can alter a variety of parameters in this area - whether for good or bad reasons. In the next section we'll provide a summary of the most essential features.

Hand control can be split into three parts at the top are the three keys: ALIGN, BACK and ENTER, in between are keys for direction using which the telescope can be moved. Below is a numeric keypad that keys also function to summon additional commands or catalogs of objects.

If you are switching on a telescope your first time you will be able to select the language you prefer using your hand controls. Enter the number of your choice language and press ENTER to select it.

The core of computer-controlled Celestron telescope is the NexStar+ hand-control

ALIGNMENT

The telescope then asks to select the alignment method you would like to use. Choose SkyAlign for telescopes with an alt-azimuth angle or Two-Star-Align to use with equatorial telescopes. these techniques usually yield the most effective results when viewing the sky at night.

Then, you will be required to enter dates, times,, and location by using the up and down keypads (6 as well as 9) on the numeric keyboard (not the big Arrow keys that can move the mount, even though it might be tempting) and the number keys as well as an ENTER button. You can edit the entry using BACK. After the first time the program will remember the place. If you need to change it then use the BACK key repeatedly to switch locations. You can either type in your coordinates or choose the closest major city from the global or national database. Even if your residence isn't listed the closest town is typically sufficient, even though the distance should be less than 100 km or 150 miles is ideal. In general, this is enough to allow an object to appear in the telescope. The software has to be aware of the coordinates to determine the time when planets and stars set or rise.

The mounts with the Real Time Clock (RTC) include a built-in clock which means you only have to input the time once and look it up in the future (in American date format!). When the mount is delivered the RTC is typically turned off to save the battery. It is turned on independently. You also have to input your time zone as well as daylight saving times manually on telescopes that have GPS receivers.

Now you are able to do the alignment according to the instructions in the chapter "First Observations" to align alt-azimut and parallactic telescopes.

What can you see?

After the alignment is successful, you are now able to begin observing. If you know what you'd like to see during the evening, open a variety of catalogues through a variety of quick keys. SOLAR SYSTEM contains the planets in the solar system, as along with the Sun (this helps with fixed mounts in the daytime however the Sun has to be activated as an object for observation). STARS provides a list of catalogs of stars (according to the SAO number, constellations and many more) as well as DEEP-SKY which includes among other things the NGC as well as also the Messier catalogue. If you are using the earlier NexStar+ versions you can also access the Messier catalogue by pressing the HELP button. Enter the numbers for the Messier objects that have three figures, i.e. using the leading zero (e.g. with leading zeros (e.g.).

The IDENTIFY keys list all objects within the vicinity of the coordinates you have specified and allows you to discover what else you are able to see through the eyepiece. ObjECT INFO gives you short information about the various objects that you've recently put up.

When you're brand novice to the hobby, particularly for newbies, the STAR TOUR button is very fascinating: once you click it the list of objects currently above the horizon will appear. Be aware that the objects listed have a horizon above them it doesn't mean they're also visible with the scope or observation conditions. The database is identical for all telescopes, and has extremely beautiful and bright targets. However, even when you're looking at a bright sky, some objects may still disappear (like numerous smaller Nebulae) or are (like many planetary Nebulae) only visible at high magnifications, or are quite massive and require a huge area of vision (like the bigger galaxies). Even if the computer offers suggestions for observation, you must not be relying on it in blind faith however, you should seek out details about the objects you can research. The OBJECT INFORMATION key provides the first clues but a good guide is always worth reading.

FURTHER SECTION

While observing during observation, when you are watching, the MOTOR SPEED key is extremely important. If you hit this key along with the number key, you will change the speed of slewing. When you set it to "9" then motors are running at their maximum speed. For instance it is possible to make use of the speed of this motor to drive towards an object that you easily see with your naked eye however the catalog number is not yet known. The slow speeds are utilized to make sure that the object is perfectly placed within the eyepiece or cameras sensors. When the speed is slower that are used, the directions are reversed so that they are in line with the view of the eyepiece as well as the viewfinder.

With MENU you are able to access more advanced options, access catalogues, and indicate if you've put two telescopes at the same time or not, and then set the slew directions and the maximum speed of slew (with "Custom Rate 9" - e.g. when the speed that is highest will be too noisy for your needs) or reset the default settings in the event there are issues. The functions are explained thoroughly in operating manual and are described in chapter Inside NexStar.

Press the BACK key to go back to the menu that you were in.

Chapter 6: The First Objectives

If you can it is possible to place your scope in bright light prior to the night you begin observation. This allows you to become familiar with all the features and gain a feel of the view that each eyepiece offers, and the way that the telescope performs when you shift one of the axes and the opposite. Explore the natural world however, be sure not to stare towards the Sun. Without a proper solar filter, you could damage your eyes forever.

Always begin your observations with your eyepiece with the lowest magnification i.e. the one with the longest focal length (the most extensive number). When using a Schmidt-Cassegrain or refractor, you must also employ the diagonal of a star; it provides a pleasant perspective and is essential on a lot of refractors to bring them in the focal point.

If you have set up your telescope prior to sunset, you can inspect the entire scope without any tension.

Do not overdo it on the first time, and be patient! Pick four or five easy goals instead of placing your self under pressure and trying to be able to see everything in the first time. If you see the Moon or bright planet is visible in the sky, begin with the brightest star. Moons that are crescent are an stunning location, since we can see that the Sun is just rising at the boundary between light and shadow mountains and the walls of crater walls create long shadows. When the moon is full there is a full moon. Sun is right above the visible portion of the Moon and with no shadows, contrasts aren't visible however the ray systems of certain craters can be clearly seen. They are composed of light-colored rock, which was thrown away by meteorites that struck. At night, when the moon is full planets and double stars are excellent targets, while less powerful objects are eclipsed due to the Moon.

The planets are tiny even with a big telescope however, clouds, as well as those of the Galilean moons that orbit Jupiter and those rings that surround Saturn along with the phases Venus are stunning to observe. Mars as well as Mercury require attentive examination and high-magnifications particularly on Mars there are plateaus and polar caps. Uranus and Neptune have less clarity and are essentially small discs. Planets appear small as compared to the Moon and sometimes are shadowed in the Moon. The next page illustrates the Moon as well as Saturn during the time of such an collision. It is important to note to note that the planets are located high in the sky. objects that are close to the horizon are more affected by the turbulences in the air.

This Crescent Moon is the ideal spot for your first night of observation. Do you see tiny Saturn just in front of the moon in this picture?
Double stars can also be good targets for telescopes with smaller apertures. In this case, you can utilize large magnifications. Double stars that have stars that are similar in magnitude are especially good to break into two. Be aware of the colours of the stars. They are one of the astronomical targets whose color is most likely to be observed.
Open star clusters generally look best when you use moderate to low magnification. You can then observe the stars that are embedded in their surroundings. The nature of the cluster is the most obvious. With higher magnification, the entire area of view is filled of stars!
Globular clusters require bigger telescopes. With an aperture of between 15 and 20 cm , you will be able to observe several pieces cotton wool that has a rough surface. With smaller telescopes, the attraction is in knowing what you're viewing.

Galaxies need a dark sky They are large and therefore, take an examination at moderate to low magnification. The darker the skyis, the more clearly you will observe it. The core of the spiral arms is the most prominent portion The spiral arms can usually only be observed in a dim light when using blindness - after all, you're looking at objects that are millions of light miles away!

Nebulae of the planets appear like faded stars with low magnification. They also exhibit their first structures when the increase in magnification.

Emission nebulae are extremely difficult objects (except in the case of Orion Nebula) which require a big aperture, dark area as well as a lower magnification. Then you will be able to glimpse their shimmering ghostly glow through the telescope.

The following is applicable to every observation Be patient. If you simply glance at the objects instead of taking a look you'll miss a lot of aspects, and accurate observation is an art of practicing. As time passes, you'll uncover many things that remain hidden for a while.

ORIENTATION in the sky

To identify the objects, look at the sky to the star charts. Find a bright star that is central in the finder, and then shift from one asterism to the next one until you get to your destination this technique is known as star hop. It's useful to draw an outline using the view fields of your finder as well as an eyepiece to use for your maps. If you are able, consider using an adjustable red light torch that dims. So that your night vision won't be impaired and you are able to identify weak targets.

HOW to DISCOVER

The eye requires 20-30 minutes to become accustomed to darkness and be able to recognize light sources that are weak. To keep your night vision intact avoid bright light sources as well as flashlights with white light.

If you are looking at objects with a lot of light that are extremely dim, you should employ a method known as indirect vision. It is not necessary to gaze directly at the object and place it on the edges in your area of vision in order to make use of the more sensitive visual sense cells located on the edges of the retina. While these provide

only an image in black and white but they also respond to lighter light stimuli.
NexStar, Alt-Azimuth - Quickstart

Once you've put together the telescope and connected the necessary accessories (finder eyepiece, a telescopic swivel, and hand control) turn the telescope on. After a short period (do do not touch any keys during the time) it will display the text "Ready" appears . The prompt prompts you to hit the ENTER button to begin the alignment. Check that the axes have been locked.

After pressing ENTER, users will be asked to choose an aligning method. Utilize the UP/DOWN buttons to select SkyAlign and then press ENTER.

You must enter or confirm the location as well as the date and time on devices that don't have GPS. Enter ENTER to confirm, or BACK to remove. Enter your personal coordinates of your location (Custom) or choose larger cities from the lists (US/International). This location is required only and is saved once you have set up your telescope first time. If you wish to change it, use BACK to go from the menu of times to the menu for location and then change it.

You should also adjust daytime saving (on/off) and the time zone for GPS devices before entering the time. devices with built-in and activated real-time clock will can remember the time, but you might need to make corrections using BACK. Be aware that the format for dates is MMDDYY.
The actual SkyAlign procedure starts. In accordance with the directions in the display, first , slew towards any star that is bright using the direction keys on the control hand. Then, place it on the finder, then confirm by pressing ENTER.

Take a look at the eyepiece and then center the star. The motor's speed instantly reduced upon pressing ENTER. To reduce the amount of gear play make use of the keys for upwards and downwards for the last movement. Make sure to confirm the alignment with the ALIGN.

Repeat the procedure with the two stars that shine as close as is possible. If the alignment succeeds, then the message will display: "Success - Press ENTER to continue". Press ENTER to finish the aligning process. When you press ENTER, the text "StarPointer off" warns you to switch off your red dots locater if you need to.

Now , the telescope is in use, and you are able to begin watching. The instructions are located on the next double page as well as in the chapter on Observing with NexStar hand control. NexStar Hand Control.

NexStar, Alt-Azimuth - The first observations

While you are able to begin watching immediately on telescopes that do not have any computer controls (if you know the location to aim it) If you are using the computer-controlled telescope, you'll need to go through an alignment process. It differs slightly between mounts with an alt-azimuth or equatorial.

To ensure that the control unit can perform its job, it is essential that the tripod is placed horizontally. Some tripods have a bubble level built-in.

In general, the following rules apply when entering numbers. Numbers are entered using the numeric keys on the hand control and menu items are chosen by the keys UP/DOWN (numeric key 6 or 9) and then confirmed by pressing ENTER.

THE BASIC START-UP

After installing your alt-azimuth mount and attaching the finder, eyepiece and, if needed the star diagonal, switch to the mount. The hand control takes couple of seconds to get going and you should not use any buttons during this period of time. If it shows something like "NexStar Ready" or the name for the mount Press ENTER to begin the alignment. Select SkyAlign by pressing the up and down buttons. You will then be asked to input the date and time , as explained in the chapter before. It is important to make use of an American dates format. For devices equipped with built-in GPS or SkySync, the alternative SkySync GPS receiver, this is automated; however, the time zone and daylight saving time have to be manually entered.

Make sure that the clamps on the axle of the mount are securely tightened and tightened if they're you have them (LCM, SLT and SE are equipped with slipping clutches, or are already clamped for CPC and Evolution, the CPC as well as Evolution both clamps on the axle must be tightened by hand). Then you can begin an actual alignment.

THE ALIGNMENT

If you choose SkyAlign the hand-control will prompt you to aim your telescope towards the first stars of alignment (i.e. align it with the finder) then confirm the alignment by pressing the keypad. Enter. Start by selecting any bright star, using only the keys on the control for hands. When you hit ENTER, the motor's speed will be reduced. When you looked into the finder, the motor was moving fast, however when you adjust it the speed will be slower.

Then, look through the eyepiece, and center the star also. When you've done this, you will be able to instantly check if the viewfinder has been properly adjusted. It is important to make the final motions of the telescope using the direction buttons on the left and down. This compensates for any gear-related play that is inevitable and

makes the mount perform more precisely. Check the position using ALIGN immediately after the star is properly centered within the eyepiece. The more precise you work this way, the more precisely the software will pinpoint the exact location.
The process is repeated for two additional stars, which you can approach once more using only the keys for direction of the control manual. The ideal is that the stars be as separated as they can and form the largest triangle.
Then, the program will calculate for a period of time, and will then display the success message (e.g. "Match" (or "Successful") after which all is well and you are now able to observe. The tracking feature that was turned off prior to (after all the telescopes do not know where the north is and how it ought to be moving to track without an effective SkyAlign) is now activated. With the BACK button you can check which stars you've used by pressing ENTER. You then complete the alignment, and you can look around.
The alignment didn't work If "Align Error" is shown. Retry the alignment, ensuring you

have entered the correct date and you have the stars as close as they can be.

The first object

You can now begin your first observation session, as shown in the illustration in the chapter on watching. If you are aware of what you wish to observe, just select the relevant category by clicking on buttons like the SOLAR SYSTEM, DEEP SKY or STARS button (often you can select a specific constellation) Select the catalog or object using the buttons UP/DOWN, and then confirm it with ENTER. If an object is displayed directly, just select it by using UP/DOWN, and ENTER. However, for other objects that have an assigned catalog number it is necessary to enter the number directly. If required, you must include leading zeros as well, e.g. 042 in M42. After confirmation using ENTER the telescope then shifts to the coordinates of the target. Check to ensure that it does not hit anything and make sure that all cables for any equipment that are connected are of sufficient length. If you encounter problems it is possible to stop the slew's motion at any point with a single press.

When the telescope has got close to the object, it'll perform a few minor corrections. The display displays a small egg timer" to the right of it in a rotating way. when it stops the object will be visible through the eyepiece. When you press the INFO button, you can look to retrieve information about larger objects that are brighter.

If you've been struck by the weather, but you haven't had the time to select objects in advance for the night, use the TOUR function. The telescope will recommend some objects that are visible. But, this suggestion is not based on the dimensions of the telescope or the impact of the ambient light, which means that certain targets might remain unnoticed even if they're (jus)t over the horizon.

To activate this feature simply hit the TOUR button, choose an object that has UP/DOWN, and press ENTER. After that, it will be automatically accessed.

Within the section "The Sky" there is also an a brief overview of all four seasons, including star maps. A few of the most beautiful locations are described in greater depth there.

A LOOK AT THE TELESCOPE

Always choose the eyepiece that has most focal length (the most extensive number that is printed onto the lens) first. It offers the most magnification, the smallest field of view , and the most bright image. It allows you to locate objects, even if they're not perfectly aligned. Be sure to be focused!

The view through the telescope is quite different from the bright images that have formed our perception of the universe prior to the Hubble telescope was launched: You'll be able to see the stars as black or white but with greater contrast. For the naked eye, nebulae are just glimmering gray areas. Begin with bright objects such as the large planets and moons, or star clusters.

Nebulae and galaxies are more demanding objects that require the darkest skies possible. Double stars are unique, as you can only see lights, but they are able to form beautiful pairs. If you focus the telescope and you can be able to see the colors of stars a bit more clearly. However, regardless of how high you magnifyit: No telescope is able to display stars in the form of balls If you've focused properly and your optics are well adjusted (collimated) stars are only pin-points. The most beautiful objects are asterisms such as open clusters of stars.

NexStar, Equatorial - Quickstart

Once the mount is aligned to the celestial pole , or close to Polaris the software needs to be initiated. Allow the hand control to start and then enter when asked.

The telescope is now prompting you to set the telescope to the marks of the index.

Then, move the axles towards the marks of index (here in the case from the CGEM) then fix the axes and confirm the clamping with ENTER.

Then you need to input or verify the date, location and time, as well as daylight saving hours (on/off) and the time zone. Enter the information and confirm or to cancel by pressing BACK.

The format for dates is MMDDYYY.

It is saved and requested after you have set up your telescope first time. If you wish to change it, go back to the menu for time to the menu for location.

Select an alignment process using the SCROLL buttons. Then confirm by pressing ENTER. Tips Use The Two Star Align for the most effective results.

The telescope indicates that it is an alignment star. Press ENTER to slew towards it. Watch until the "egg timer" disappears from the screen. Place the star in the eyepiece and the viewfinder Make sure to confirm it by pressing the ENTER key and then align. A majority of aligned stars are displayed on the charts for stars within this guide.

Repeat the process for a second star and answer the question for a third star using BACK (or add two stars to improve accuracy). It is available to be used. The functions are available via in the menu main (CGEM: CGEM ready).

POLAR ASSIGNMENT TO THE ALL STAR POLAR ALIGN

If you've polar-aligned the mount, but only in a rough manner then, choose one of the stars from the STARS catalog that is half way towards the southern. The star maps within

The Sky chapter" contain names for some of the more prominent stars. Alternately, you could navigate to a star using the direction keys, and have it recognized by pressing an identifier key. Then press ENTER to move toward it.

Press BACK for a return to main menu, and hit the ALIGN key. Use your SCROLL keys to pick the Polar Alignment after which you can select Align Mount. Follow the steps on the control for hands. The mount will then move to where the star would be in the case of a mount that is aligned. When you have a high magnification, it is important to place it in the eyepiece by moving all the mount's components using your tilt (right/left) and the altitude setting screws. Don't use the hand controls! Make sure to enter ENTER and the mount has been aligned polar. Turn it off , then back on in order to reset your reference star and make an alignment. You can now observe. For more information, continue to the next page. You will also find observation using hand controls like the NexStar hands-control.

Chapter 7: Nexstar Equatorial The First Observations

Although you can begin observing straight away using the telescope that is not controlled by computers (at at least, in the event that you know where you want to put it) when you use an electronic telescope, you'll first need to go through an aligning process. The alignment procedure is different between equatorial and alt-azimuth mounts.

In order for the computer's control to function properly the tripod needs to be set up horizontally (some mounts have a bubble level built-in). Additionally, you must to align the polar axis as precisely as you can, as detailed within the section "Setting the Equatorial Mount". For observations using visuals A fairly rough alignment is adequate to observe Polaris via the polar scoper finder, telescope or. The hand control is able to adjust for an inclination when the object is positioned, but the star will gradually disappear from the field of view with time.

In general, this is the case to numbers: They are entered using the numeric keys on the hand control and menu items are selected using the keys UP/DOWN (numeric key 6 or 9) and then confirmed using ENTER.

THE BASIC START-UP

After installing your equatorial mount (Advanced VX CGEM II CGX(L) or CGE Pro) and attaching the eyepiece, finder and, if required the star diagonal, turn to the mount. The hand control will take just a few seconds to start Do not press any keys during this period. If "NexStar Ready" or something like that is displayed on the screen, you can press ENTER to begin alignment.

After pressing ENTER You'll be asked to shift the mount's axles in such a way that the marks on the index coincide (Set for Index) This ensures that your telescope will be placed according to the image within the section "Polar Alignment" The telescope is located on high on the mounting and is pointed toward the pole of celestial light.

Then, you will be asked to input the date and the time as in the instructions in the page. Make sure you are using your

American dates format! If you have a device with built-in GPS or SkySync, the alternative SkySync GPS receiver, this happens automatically. However, the time zone and daylight saving time have to be set manually.

THE ALIGNMENT

The actual alignment will begin. Select Two-Star Align as an alignment method. The control on your hand asks you to select an option from the list. You can confirm it by pressing Enter. It should appear within the skies. The telescope will rotate towards it in a way, but you'll be required to manually center it. The charts of stars contain the most significant stars that have names. Be assured that the star isn't located directly within the finder, although it could occur. If it isn't centered then center it within the locater (the display will read "Center [Star Name[Star Name]") with only the control buttons on your hand. Once you hit ENTER, the motor's speed will be reduced If, when you used the locater, they were going rapidly, but for the next small adjustment, the speed is decreased.

Then, look through the eyepiece and then center the star there too While doing this, you can instantly check if the viewfinder is properly adjusted. Important: Make your last motions of the telescope last using the direction buttons upwards and downwards to compensate for the inevitable gear play and makes the mount operate more precisely. Make sure to confirm with ALIGN as quickly that the star is located within the eyepiece. The more precise you can do this, the better the software will be able to determine the exact location.

This process is repeated for a different star, and you center again using the directions buttons of your manual controls. The mount will suggest an additional alignment star and vice versa.

The mount asks you whether you'd like to add a second calibration star. By pressing BACK you can complete the alignment or can add additional alignment stars using ENTER, which will further enhance the GoTo.

It is now recommended to carry out the All-Star Polar Align as described earlier by accessing the main menu by pressing the

BACK button, then pressing ALIGN, then select Pole Alignment or All Star Polar Align (depending on the firmware version). You can then check the deviation or execute the exact Polar Alignment. If you decide to do this then you will need to repeat the two-star alignment once more.

The first object

You can now begin the first observation. It is shown in the illustration in the chapter on watching using your NexStar Hand Control. If you have an idea of what you wish to observe, just select the appropriate category using the SOLAR SYSTEM OR STARS, Deep Sky keys (often you can sort by a particular constellation) Select the object or catalog using up/down and confirm by pressing ENTER. If the object is listed in the catalog, you can choose it using UP/DOWN and then ENTER. For objects that have a catalog number you have to enter this number in the field with the number. If required, you must also type in leading zeros, e.g. 042 to indicate M42. After verifying by pressing ENTER the telescope instantly moves to the coordinates of the object. Be sure to not touch anything or ensure that any cables from accessory connected to it are sufficiently long. If you encounter problems you are unable to stop the move at any time with a single press.

When the telescope has reached the object after which it makes some adjustments. The display displays a tiny egg timer" on the right side as moving line. When it's stationary, the target is visible through the eyepiece. By pressing the INFO button, you can look to retrieve information on more bright locations.

If you've been awestruck by the beautiful weather but did not have time to choose which objects you would like to see the night, make use of the TOUR function. The telescope can suggest some objects that are visible. But, this suggestion is not based on the size of your telescope , or the effect of light pollution. Therefore, certain targets might remain unnoticed even though they're (just) just) above the horizon.

To access this feature you just need to hit on the TOUR button, choose an option with UP/DOWN and press ENTER. After that, it will be accessed automatically.

The chapter "The Sky" you'll also find an information on the four seasons, including star maps. A few of the most stunning objects are described in greater detail.

A LOOK AT THE TELESCOPE

Always opt for the eyepiece with most focal length (the most extensive number that is printed onto the lens) first. It provides the lowest magnification, with the biggest field of view , and the most bright image. This lets you find objects even when they're not perfectly aligned. Make sure you concentrate!

The view through the telescope is different from the vibrant images that have formed our perception of the universe prior to the Hubble telescope can be seen with a black-and-white color however, with a greater contrast range. For the naked eye, nebulae are always glistening gray areas. Begin with bright objects such as the moon, the planets and star clusters.

Nebulae and Galaxies are the most demanding objects that require the darkest skies possible. Double stars are different, since here there are only the light points, however they are able to form beautiful pairs. If you focus the telescope it will allow you to observe the colors of stars more clearly. But regardless of how much you magnifyit, the telescope cannot display stars in the form of balls If you've focused properly and your optics are collimated, they are simply pin-points. Most appealing are asterisms that resemble open clusters of stars.

Utilizing using the NexStar Hand Control

After alignment that is successful, i.e. the initialization of the telescope, it is now operational. The entire operation are controlled with the hand controls. It's divided into three parts at the top are the three buttons required to align of the hand control: ALIGN, ENTER and BACK. BACK brings your back into the main menu.

Below are the keys for direction that are arranged in an oval that lets you can move the telescope once the axes are in place. Don't loosen the axes while you are watching as the computer will not be aware of what direction the telescope is. When you press one of these buttons, it can also stop a slewing motion. This is crucial in the event that, for instance, the object is automatically positioned and the telescope may be struck by something.

With the numeric keypad you can browse through various catalogs, and then select the appropriate object. You can also choose the speed by using MOTOR SPEED. Numeric keypad "9" is the fastest mode.

The mount's settings can be altered through MENU (numeric number "7"). For more details, refer to the instructions for the mount and inside the section called Inside NexStar.

The STAR TOUR button opens an option to tour. The telescope then suggests the most stunning objects visible. Click SCROLL UP/DOWN for scrolling through the list, then press ENTER to switch the telescope towards the desired object.

By using the DEEP Sky key, you can open a variety of catalogs of deep-sky things, i.e. objects that are not part of our solar system. Scroll through the list of objects, then hit ENTER to choose the catalog.

Then, input the number of an objects (with preceding zeros e.g. 0001 for Messier) and then navigate towards it using ENTER.

Its SOLAR SYSTEM key calls up an inventory of all planets that are visible within the skies. Only the Sun is required to be activated in isolation, since it requires a filter that is specifically designed to enable it. With the help of SCROLL UP/DOWN, you can browse through the listof objects, and by pressing ENTER you choose an object. It can then be accessed instantly.

There's a brief information text for many objects which you can search for by pressing the OBJECT INFO key.

The Sky

In the modern age of smartphones with planetarium apps it's worth taking the time to be aware of the sky. The following pages are going to find stunning objects of each season, as well as star maps that cover the entire year, a tiny lunar map for the beginner and some tips to observe the Sun and the variable stars.

The maps of the stars contain initial targets for observation; for more targets and descriptions, consult other sources that will cover the subject in depth or you can use the tour mode feature of NexStar. NexStar control.

The Moon
The Northern Sky
The Spring Sky
The Summer Sky
The Autumn Sky
The Winter Sky
Star Maps
Watching the Sun
Variable Stars

The Moon

Through the naked eye, we are able to observe that the Moon changes constantly when it moves around the Earth. Since the line that divides night and day (the "terminator") continuously shifts across the lunar surface New mountains and craters appear every hour from the darkness, whereas others are pushed further towards the bright sunlight which means that their shadows get shorter. The most fascinating images are those of the Crescent Moon and the area near the terminator. At the full Moon and beyond, with the exception of brilliant ray systems of a few craters , such as Copernicus or Kepler there aren't many structures are visible.

A telescope can show structures as small as a dozen kilometres and allows you to explore walls, gorges, and craters that our satellite. The Moon is the only celestial object that has clear structures like the Moon. Moon.

The craters are remnants of meteorite collisions and are found primarily in the highland regions with bright light. They are among the oldest regions on the Moon and their rock formations are around 4 billion years old. The biggest craters are more than 100 kilometers in diameter The smallest is visible with the telescope, only several kilometers.

The dark regions - maria maria, plural maria, that refers to seas, are massive plains of lava. This is where the crust was broken through meteorite impacts and lava filled the enormous impact craters. They created during the most recent heavy meteorite blast around 3.8 billion years back. They are therefore younger than plateaus with crater-filled craters, which is why they have very few impact craters. Following the last major blast, it has remained tranquil within the Solar System. When the lava cools it turns to basalt consequently the maria of Moon are very dark. In certain places, you are able to see the older crusts were submerged or were cooled by waves.

If you're looking at the Moon make sure to ensure that it's at the highest point on the sky as you can. If it's near the edge of the sky line, air turbulence will disrupt the image, and you'll only be able to observe only a few small tiny details. If it's about halfway to the sky's top it is possible to magnify to the maximum that your telescope can handle as well as the air-turbulence makes it possible. It also makes the image darker but it's beneficial to utilize moon filters, particularly for those with low magnifications. This kind of neutral density filter is put to the optic. The popular lunar filters in green are especially appropriate for achromatic lens telescopes that have a certain color aberration. They reduce the fringes of color and allow sharper images as they are designed for the colors that the achromatic telescopes have the sharpest images. The downside of this is the fact that moonlight gets tinted with green.

With apochromatic refractors of high-quality, Newtonian mirror telescopes or Schmidt-Cassegrains with neutral or gray density filter is more suitable since it doesn't distort the color. A adjustable polarizing filter is useful: in this case two filters are twisted in opposition to one another and can alter the brightness of the filter. This is especially useful when you can put an individual polarizing filter into the eyepiece, and another to the diagonal of the star or an 1.25" reducer. Then you can change the brightness simply by rotating the eyepiece.

The moon's appearance is stunning using a binoviewer which you can view the moon the moon with both eyes. Then the craters appear to be three-dimensional.

This map located on the following page provides a brief look at some of the craters.

If you'd like to explore further, you must search for a lunar atlas that is large.

Mondmeere

(A) Mare Crisium

B) Mare Fecunditatis

C) Mare Frigoris

D) Mare Humorum

E) Mare Imbrium

F) Mare Nectaris

G) Mare Nubium

H) Mare Serenitatis

I) Mare Tranquillitatis

J) Mare Vaporum

K) Oceanus Procellarum Krater

1.) Archimedes

2.) Aristoteles

3.) Aristarchus

4) Aristillus

5) Atlas

6) Copernicus

7) Deslandres

8) Endymion

9) Eratosthenes

10) Eudoxus

11) Gassendi

12) Grimaldi

13) Kepler
14) Langrenus
15) Plato
16) Posidonius
17) Taruntius
18) Theophilus
19) Tycho Gebirge
A) Montes Alpes and Vallis Alpes
b) Montes Apenninus
c) Montes Carpatus
d) Montes Caucasus

Chapter 8: The Northern Sky

The northern stars in the sky always remain identical, but their position changes. Since the celestial pole never moves precisely in the north, and the stars revolve around it, the stars closer to the horizon will do not change direction. When using long-exposure photography, in which the camera is pointed towards north-facing skylines for a lengthy duration with no tracking could be clearly seen.

If you expose a photograph that shows the sky a few days, stars form an arc around the pole of the night.

The most well-known Northerly constellation can be identified as known as the Big Dipper. It's not really an actual constellation since there are several other stars that are part of this constellation, which is officially known as Ursa Major, also known as the Great Bear, but these seven stars are easy to spot and are well-known by the name of Big Dipper, and under various other names across other countries. It's an excellent way to find your way when you expand the line of the front star boxes by five, you'll arrive at the pole star. And If you follow the curve of its handle during spring it will reveal its brighter stars Arctur within Bootes along with Spica within Virgo.

The central star in its handle makes a great initial target for telescopes. Mizar will be the more bright star and the companion star Alcor is easily visible with the naked eye. Both stars are approximately 1/3 of the total moon's diameter, which is equivalent to a distance of three light-years between them. This means they're a bit more distant to each other than that of the Sun and its nearest neighbor, Alpha Centauri, 4.3 light years away. Through our telescope Mizar which is located 78 light years from us, is separated as an extremely small double star. There are 4 stars. Mizar with a tight pair of star, Alcor as a rather distant companion as well as a fourth, weaker star, that is a triangular structure together with Mizar and Alcor but is far from the pair of stars.

M82 (above as an elongated in the upper left corner as a narrow) M82 (above as a narrow line) and M 81.1 (at the lower right of the photo) are lovely pairs of galaxies that are part of the Big Dipper. The photo is similar to the view from a larger telescope that has a 20 cm (8") opening with a low-magnification.
With a slightly bigger telescope it is possible to search for the two galaxies M81 & M82, when you are looking at the Big Dipper is high in the sky. If they are in a dark spot, they can be observed with an aperture of just 80 millimeters. They appear as tiny Nebulae when in good lighting conditions however, you'll require a more powerful telescope to observe the structures of the two galaxiesthat are approximately 7.2 or 18 million light years from each other. M 81 is one of the most bright spiral galaxies. We

view it from the top of the sky from above. Check to see if are able to recognize other structures through the indirect eye. M 80 is an irregular galaxies and we can see the galaxy as an elongated line which displays first structures with higher magnification, or in photos.

Cassiopeia's constellation is often referred to as "Celestial W" Five stars are visible to each other to the Big Dipper as a big M or W in the sky, based on the time of year, and it is the Little Dipper with the pole star is located between them.

The Milky Way runs through Cassiopeia which is why we have many open star clusters in this region. The most stunning is M52, which has nearly 200 stars. The cluster, which is at 3900 light years distant, has the diameter of 14 light years . It can be separated into numerous separate stars with the telescope. M52's adiameter is a half of the moon's full size and in the center, the stars are packed.

It is believed that the Little Dipper is especially famous for Polaris the pole star that is the point at which it reaches the end the handle. It is brilliant, but it's not the most visible night sky star. The constellation is sadly, however, it contains none of the other objects that are visible.

Between the Big the Big Dipper and Little Dipper Between Big and Little Dipper, the obscure Draco is the Dragon is a solitary star that flies throughout the skies. It is Draconis (Kuma) is among the stunning double stars that small telescopes can see and is the symbol of the Dragon. Double stars, located 100 kilometers away is made up of two white stars and is visible through an eye that is naked as one star. Their position hasn't changed since 1833, however it's still an actual double star. they orbit each other similar to how the Earth revolves around the Sun however at an increased distance.

This Cat's Eye Nebula NGC 6543 is a small planetary nebula. It is seen at lower magnifications as either a blue or greenish-colored star that lies between d and Draconis. The nebula is 3000 light years away is the remains of a star that went out of existence around 1000 years ago. The nebula is today approximately one-third of a light year's the size. In larger telescopes, it transforms from a light spot to a tiny disc and with a bit of imagination, it can appear to look like an eye of a cat.

Chapter 9: The Spring Sky

In spring it is when spring is when the Milky Way is quite close to the horizon. This means we are able to look beyond the horizon of our home galaxy, and out into the deep space. Three bright stars make up the spring triangle, also known as Regulus in Leo which is also known as The Lion; Spica in Virgo The Virgin; and Arcturus in Bootes The Herdsman.

Spica is by far the most bright star in the large, but obscure Virgo. The "realm of galaxies" is situated within Virgo along with the Coma Berenices. Berenice's Hair It is located in the middle of the Virgo galaxy cluster, approximately 65 million lightyears distant. In an area of size of 16 million light-years there are approximately 3000 galaxies. They are part of a larger supercluster that includes the groups of galaxies in Leo and the Big Dipper as well as our personal Milky Way. Numerous galaxies in Virgo are visible through an aperture of eight or 10 centimeters (3 or 4 inches) at the very least from a dark spot. For spiral

structures, but a 20cm (8") apertures is recommended.
One interesting object within this region is M64 The Black Eye Galaxy. The dust bands named for it can be seen only with telescopes that have an aperture of at least 150 millimeters. The spiral arms tightly wound as well as a huge cloud of dust will be visible. The smaller optics can only reveal the appearance of a tiny, fuzzy star. Its diameter is 65 000 light years. M64 is the most bright galaxy in Coma Berenice as well as one of the brightest 12 spiral galaxies visible in the sky.

The two galaxy M65 and M66 appears as blurred star formations in Leo the Lion The thin line of the galaxy NGC 3628 is also seen

in the upper right. The view is comparable to the view of a larger telescope with a low magnification.
The group of galaxies within Leo are close to us as well as the Virgo cluster and are able to be located even without a GoTo system: The couple of M65 and 66 are slightly further south than Theta Leonis which is a prominent star in the fascinating constellation Leo. M65 is located 56 million light years away, while M66 is just 32 million light-years away. The bright central region of M66 is easy to locate, while the hazy glow in its spiral arms is visible with low-magnification. M65 is a more dark version of the dusty M66. It is situated close to it by about two thirds of a complete moon's in diameter. The north of the pair is the smaller galaxy NGC 3628. We can directly observe its edge. These galaxies first came to light in 1780. Their centers will be visible as blurred stars in the dark skies.
Gamma Leonis (Algieba, ,,forehead") is located in the middle of the head and body of Leo which means that Leo's constellation Lion is looking directly at us. The small pair of stars with different brightnesses could

easily be distinguished at greater magnifications with the telescope. Both stars glow yellowish even if the smaller one is described as green. They require 618.5 years to travel around the common center of gravity.
In Canes Venatici in the home of the Hunting Dogs is an iconic Whirlpool galaxy M51. It's a gorgeous pair of spiral galaxies which collided over 56 million years back. They remain a reminder of this meeting today. Because they are frequently photographed, they might not be as impressive when you view them by a telescope first time around: The spiral arms are identified as ghostly glows on smaller telescopes, however they're not as striking like the well-known images. But, the couple is always worth a look.

Globular star clusters are able to be easily observed with an aperture of around 20 centimeters/8". In smaller telescopes they appear like cotton balls, and might appear a bit grainy on the edges. M3 is located within the Hunting Dogs in the Bootes and also the brilliant Arcturus located in Bootes. It is among the three most bright globular star clusters visible in the sky. Only M13 is the brightest. Hercules is more bigger and more brighter, but M13 is much lower than the rest of our sky. M3 remains with Canes Venatici The Hunting Dogs, but it is near the boundary to the more prominent Bootes. The 45,000 stars it has scattered over a half the circumference of the moon's full size though it appears to be much smaller when viewed through the eyepiece. the size of its area is only evident in photographs. At 34,000 light-years away that is roughly 170 light years in size. As with all globular clusters of stars they are extremely old, with a total of 10 billion years old, which is more than twice the age of our sun.

A distinctive "ice-cone" model of Bootes The Herdsman is easily visible as is its bright main star Arcturus is easily located when you follow the turn on the handle of the Big Dipper. This constellation has some stunning double stars.

Delta (d) Bootis is an exquisite, large pair of stars that was first observed through F. G. W. Struve in 1835. Since the second star is weaker that the primary star, and can easily be outshoned, you may want to use a greater magnification. About 120 light years distance us from the gorgeous couple.

My (u) Bootis is sometimes referred to as Alkalurops. It is an amazing triple star. When you look with the naked eye, you can see a blueish-white but not an incredibly prominent star. Even binoculars show another star that is not apparent to the naked eyes. When magnifying it higher, another companion is visible, just 43 astronomical units distant from it. One AU equals the distance Earth-Sun , or around 150 million kilometers. This means their distance from each other is a bit more than Pluto's distance to the Sun. The two stars rotate around each other once every about 260 years, forming an actual double star that is that is 121 light years distant. The brighter star in the trio is located 3000 astronomical units from the two stars.

The Summer Sky

In the summer during summer, during summer nights, the Milky Way stretches high above our heads. Unfortunately, these days, you must travel quite a distance to be able to appreciate its splendor. It is often

overshadowed by the lights in our urban areas.
The spring constellations can be seen to the west but the three stars which will be visible first at sunset (together along with Arcturus and Arcturus in Bootes) create the striking summer triangle. These include Vega located in Lyra, Altair (also: Atair) in Aquila and Deneb located in Cygnus. Vega as well as Altair are close neighbors of the Sun with only 17 and 25 light years apart in each case, and Deneb is over 1800 light years away. It is among the brightest stars in the universe.
The globular M13 cluster in Hercules is easily found once you've found the subtle constellation located about a third in between the two stars that make up its trapezium that is displaced. It was first discovered in 1714. It is the most beautiful star cluster that is globular in the northern sphere of the celestial sphere. In its outer regions, the first stars are located in smaller telescopes, larger telescopes may even detect it to the middle. In an area that has an area of 160 light years around 500 million of them are crowded together. This

is around 30, 000 light years away from Earth which is about as far away from the central region in our universe as the sun is. Make use of higher magnifications in order to see the galaxy better.

The constellation Cygnus is known as the Swan is situated at the center of the Milky Way and has much to provide. Its head Swan is situated in the middle of summer's triangle, and is among the most stunning twin stars. Albireo (b Cygni) is visible at approximately 100 times magnification . It is two distinct and bright stars. The color contrast between two stars is stunning: one is clearly yellow while the darker one is clear blue. It is possible to see the colors more clearly by focusing the image a bit. The distance is 380 light-years from the two, and belongs to the handful of celestial objects that clearly show some form of colour.

The open cluster of stars M39 appears best when viewing with low magnification. the brighter 25 stars that comprise the loose cluster of stars 830 light years distant appear to be the shape of a cluster. The stars make up a large triangle. The radius of the 200 million-year oldest galaxy of stars measures 7.5 Light years which makes it appear nearly as big as the full moon.

M57 The most well-known Ring Nebula in Lyra, is located between two lower corners in the parallelogram Lyre and can be seen in the upper right corner of his photograph.
A single of the more interesting objects visible in our summer skies is M57 The Ring Nebula in Lyra, the Lyre. It lies somewhere between the northern and southern corner star in the Lyre parallelogram and lies around 1800 light years from us. M57 is among the most famous planetary nebulae. It clearly displays its ring-shaped shape when magnified 100 times. In smaller telescopes it is a tiny disc, or maybe the shape of a star. This is how the scientists felt that they were reminded of the planet.
Its size is less than half a year in size, it expands at a rate of 18 km per second. Once the gas masses of its nebula have moved away from the star's remnant which causes their glow, they will be inaccessible to us. This is a fact that dates back to the demise of a star around 20 000 years ago. Its central

star remains unobserved by the majority of amateur telescopes.

Another intriguing planetary nebula can be found in one called the Dumbbell Nebula M27 in Vulpecula which is the Fox which is located about 1,000 light years away us. With an estimated diameter of 1.2 light years and a diameter of at least 1.2 light years, it is one of the biggest planet-wide nebulae. With a lifespan of 48,000 years, M27 is twice as long as the rest of the models of this type. M27 is composed of many gas sheaths, which are spread at different speeds. At low magnifications, it appears as an oval, but it is a tiny and inconspicuous tiny light spot. When magnified higher its dumbbell-shaped shape and subtle coloration slowly appear.

M27 can be observed with binoculars or in a viewfinder that has a 50mm aperture. It is easy to become confused with M71 within the constellation Sagitta M71, also known as the Arrow M71, which is approximately eight degrees, or the full moon's diameter. further to the south. If you zoom in higher, however M71 is able to be reconstructed into stars, and it is easy to differentiate. The 14,300 light-years remote star cluster likely one of the globular clusters, even though it may be considered an open cluster. The estimated age range of 9 to 10 billion years is a sign of the existence of a (relatively young) huge globular cluster.

If you have a clear view of the south, make sure you do not overlook these stars Sagittarius (the Archer) and Scorpius (the Scorpion), that only appear across the horizon for a short period of time. This region also contains the central region of the Milky Way, which remains unnoticed due to the dense dust clouds that make up the Milky Way.

The large globular cluster M22 that is just 10000 light-years away is worth looking into with higher magnification. the first stars are beginning to be observed in its fringe areas. As with M13, M22 contains around one million star clusters. It also is bigger than its counterpart in Hercules. It is, however, not far over the horizon which makes it an easy to hit. Its diameter is 75 light years, which is one-third of the size of M13. Only o

Centauri as well as 47 Tucanae seem brighter and bigger in the sky than M13.

The Autumn Sky

Similar to spring, we take a look at the Milky Way again in autumn however this time we look out into the sky in an area that is almost devoid of galaxies. The most common constellation of this time of year will be that of the Square of Pegasus formed by Pegasus and Andromeda. Check out the colors of the stars. Can you detect a hint of colour within the 4 stars?

The Andromeda galaxy is visible in the sky, six times bigger than the moon's full size - not an easy to target. The two galaxies that surround it are seen as tiny Nebulae.

The most prominent feature of the sky in autumn is obviously the next major galaxy in our galaxy: M31, the Andromeda Galaxy. It is estimated that 2.5 million light-years are separated from the biggest galaxy within our local group. This galaxy is, having a diameter of 150 000 light-years, bigger than our home Milky Way. Its bright central region is about 12,000 light-years, roughly two thirds of the apparent size of the moon's full size - all of the galaxy is to be six times as big in the sky in comparison to the total moon. In good light it can be seen by anyone with a naked-eye view as tiny Nebula and is easily visible in binoculars. With the telescope however it's a tough object: very few telescopes offer the field of view greater than 3 degrees, which means that even at the smallest magnification, only the bright, non-structureless core region can fit into the eyepiece. The two galaxies that are associated, M32 and M110, are much easier to spot through the telescope since they are extremely large and bright. Avoid high magnifications.

The beautiful two-star Gamma Andromedae (Alamak or Almaak) can be observed with an unaided eye. It appears as one star. With higher magnification, it appears like a miniature Albireo in Cygnus as the blueish and yellowish hue of the two stars is clear. The star, located about 120 light-years away is located at the bottom of the star chain in Andromeda. Try magnifications of 100x or more.

In the constellation of Pegasus in the constellation Pegasus, the M15 globular star cluster is worth a glance. It is located about a degree to the northwest e Pegasi and is almost as bright as an unaided eye to see. The compact globular cluster of 160 light years , was first discovered around 1746, through Jean-Dominque Maraldi. You will require a telescope that has at least 15 cm of aperture to be able to discern the initial individual stellar cluster of M 15, which is 30600 light years away. in smaller telescopes , it appears like a tiny cotton ball or washed-out star. The numerous stars that can be seen in the cluster are not suitable targets for amateur telescopes with a common scope. One of the most notable features in M 15 is that it was the first globular cluster which it was discovered that a planetary nebula in a photograph made at Mount Wilson Observatory. Mount Wilson Observatory in 1927.

NGC 7331 is the most bright galaxies in Pegasus and is bright in its center; however, its brightness diminishes rapidly toward the edges. The galaxy's large spiral is similar to M 31, but is about 20-30 times farther from. The distances reported in the literature range from 40 to 66 million light years. NGC

7331 is commonly used as a model to represent our own galaxy. It is viewed obliquely from above. Its center should be seen in the larger scopes. It appears to be blurred stars. Use indirect vision to observe the sparkle in the arms spiralling.

The open cluster of stars M34 is located on the border between Andromeda as well as Perseus. With a diameter that is the size of a full moon, it does not require a large magnification in order to view it. It is home to around 80 stars, and they are scattered. It is only in the center that is a smaller circular collection of around 20 stars. The distance between us and it is 1.400 light years and the diameter of its ring is approximately fifteen light years. The central region with a half dozen brighter stars is approximately four light years, which means it is exactly in within the space between us Sun and its closest neighbor which is the Centauri. M 34 is about 180 million years old.

H & Chi Persei is an unusual title for a beautiful double star cluster. It is the combination consisting of NGC 869 as well as NGC 884, both consisting of approximately 300 stars within an area bigger than that of the moon's full. With a distance of around 7000 light years, that equates to an area of around 60 light years which makes them among the brightest and largest examples of their type. With the dense galaxy field within which they're embedded the look more striking. NGC 869 is 5.6 million years old, while its neighbor is just 3.1 million years old. Around 7,000 light years separate us from the two open clusters of stars that join to form the massive Perseus Ob1 star field the name given to the Perseus spiral part that is part of the galaxy. With this huge star field in our sky the two star clusters , as well as their surrounding areas are destined to be used in low magnifications as well as short long focal length telescopes.

Algol (b Persei) is the "Demon Star of Perseus" and also gives its name to an entire set of variable stars. Two stars have a mutual orbit at the distance of approximately 93 light-years from Earth and their orbital planes being in our view. This means that the larger but weaker part of the double star system is eclipsed by the smaller but more bright main star each 2.87 days. In the course of the eclipse that occurs, the overall brightness of the system diminishes significantly for the duration of ten hours. smaller decreases in brightness are observed when the star with the highest brightness is in the shadow of the main star. The changes in brightness are visible through the naked eye as you look at the brightness of Algol versus other stars in the region. But even a telescope of this size can't resolve the two stars because just 10 million kilometers from each other.

The Winter Sky

The winter nights are illuminated by the Milky Way and the stars of the winter hexagon: Sirius in Canis Major (the Greater Dog), Procyon in Canis Minor (the Minor Dog), Pollux in Gemini (the Twins), Capella in Auriga (the Charioteer), Aldebaran in Taurus (the Bull), and Rigel in Orion (the Hunter). A massive hexagon with the middle being where Betelgeuze in Orion sparkles in red.

It is believed that the Orion Nebula is an active region of star formation and is located beneath those belt stars in Orion.
It is believed that the Orion Nebula (M42/M43) is the most bright part of a huge cloud of hydrogen, a few light years across and spanning every constellation. Today new stars are formed within it as regions of the gas cloud collapsing under their own gravity. The pressure eventually rises to a point that nuclear fusion takes place with the formation of a brand new star is ignited." The Nebula, which is between 1,300 and 1900 light-years from is still large enough to host the formation of 10,000 stars that resemble suns. The most recent version of the stars has a lifespan of just 100,000 years old, and can only be observed by infrared radiation. The stars that have been uncovered from gas clouds, are around 1 million years old. One of these is the trapezium star Theta Orionis that is visible at lower magnification as a single star within the Nebula complex. With a lifespan of around 300,000 years, it is among the stars with younger age that we can observe in Orion.

The spectacular open-sky cluster of stars in the Pleiades (M45) is a result of the same nebula complex but it's already more than 78 million years old comprising 300 stars scattered over an area of a radius forty light years. The sky in which it is visible appears twice as big in the night sky as the total moon. With higher magnifications it is possible to outstrip the scope that your scope can cover. The nebula with a shimmering blue hue that surrounds a few Pleiades stars is particularly apparent on photographs. The shimmer isn't the remains of their ancestor of the nebula but rather the dust nebula that the cluster of stars are moving. With space, this nebula appears to be difficult to target, but Pleiades itself are easy targets. Pleiades themselves are simple to spot by your naked eyes. With a distance of about 360 light years they are the fourth closest star cluster. Just the star clusters of Big Dipper (70 light years) and The Hyades (150 light years) and Coma Star Cluster (260 light years). Coma Star Cluster (260 light years) are more close to us.

Hyades Hyades is the most stunning star clusters located in Taurus and are easily found close to that bright star Aldebaran. However, this star is not part of The Hyades: Aldebaran is only 65 light years from Earth and the Hyades are approximately 150 light years away. Around 130 bright stars scatter across 60 light years. when you look through the telescope, they are a part of the image field . Take the time to look them up in the finder if you own an optical finderscope in the telescope or better with binoculars!

Although it is true that the Orion Nebula is a star-forming region, as well as its companions, the Pleiades as well as the Hyades contain youthful stars, the M1 is the end of the lifespan of a star. Crab Nebula is the remnant of a star that burst into flames during the form of a supernova. Chinese astronomers saw this explosion on the 4th of July 1054 in the year 1054 and in 1731, the Nebula was first seen by an observatory near the site where the blast occurred. In the outermost layers of this former star that make up the visible nebula, continue expanding at 1,500 km per second. The remainder of the star sunk into a dense neutron star that measures only 10 kilometers in size, and lies at the center of today's Nebula. It is now just a tiny spot of light close to Zeta Tauri - this star is one of the two horns on the bull. It's a fairly small nebula, even at larger magnification.

The M35 star in Gemini is a brightand open star clusterthat is about 2,600 light years distant with around 200 members. The brightest stars in the cluster have around 400x the brightness as our Sun. Its size is 40 light-years, the M35 is nearly as big as the moon at full. It's nearly circular, and has no star cluster worth noting in its center, even with a telescope. Be careful not to magnify it too much to give the impression of a star cluster.

Praesepe or Praesepe, also known as the Beehive (M44) within Cancer the Crab is awe-inspiring Its total brightness of the cluster of stars is more than one of the most bright stars in the constellation! With three full moons and a dazzling appearance in the scope as a field of stars. At 525 light-years away the open star cluster is comprised of around 200 members and the diameter is 15 light-years. At a time of between 650 and 700 millions of years old, Praesepe can be quite ancient.

M67 is a collection of 500 stars located far in it's Milky Way plain. It is among the oldest open clusters of stars and is estimated to be between 3.9 and 3.9 billion years. The brightest stars in the cluster are small, but the brighter 75 stars are within the reach of smaller telescopes. At the distance of 2,500 light years its 12 light years is the full moon's diameter. The stars of the constellation make up an inverted S which reminds viewers of the King cobra.

Just a bit to the north from Sirius to The Greater Dog lies M41, that was described by the year 325 BC by Aristotle as an "cloudy area." Today , a telescope shows between 50 and 100 stars over an area that is full moon size. Make sure to wait until it is at its highest in the south to ensure that it doesn't disappear into the mists of the horizon. Its diameter is of 24 light years , and is located 2,250 light years from us. It's worth taking a close look at M41. The stars appear to be forming a variety of lines and groups. If you sketch a tiny drawing of the area, you'll be able to improve your observation skills paying particular attention to the details you could otherwise miss.

Chapter 10: Star Maps

These star maps depict the sky during the night hours of the various seasons. The table that is above the maps lists the dates they are applicable - in standard time as well as daylight savings time. Map I, for example is available at 9pm in mid-January, and at 1am in the middle of November. The background color also provides an indication of when it's the day and when it's night.

The star maps highlight some of the most brilliant night sky stars that are able to be seen with smaller telescopes, provided the sky is sufficiently dark. Also, you will find the proper names of a few more bright stars. These could also be able to use as reference stars for the aligning for NexStar computer control. NexStar Computer Control.

As compared the normal maps the east and west directions are reversed. The cardinal points are also correct when you put the star map above your head. Since this may be uncomfortable over the long term it is advised to keep your eyes in the direction you are looking down. This will help you to

position your eyes better to the sky. The central part of the map is the zenith, or the spot that is directly above you.
The legend on the left side of maps will always show how bright the stars are. Stars that are 0th magnitude are extremely bright and stars with a magnitude of 4 can be easily identified in rural areas , but could be a problem when in urban areas. The legend to the right side shows which deep sky objects can be found behind the symbol.
The lines of the constellations correspond to the conventions used in Europe However, they don't have any binding. Don't be surprised when you find a constellation in a different atlas of celestials with a slightly different layout The boundaries of the constellations are legally defined, but not the secondary lines. Even the names of constellations weren't determined globally until the early 20th century. Prior to that, anyone could create their own constellations at any time.
For objects that aren't as well-known, you require a more detailed star map, which only shows the specific constellations or sections in the night sky. However, you can

locate more bright objects by using this overview of maps.

Learn to recognize the more bright constellations. Then it will be simpler to locate your way through the sky. Learn to understand the sky piece-by-piece and you'll be able to master the sky by practicing even without the map of stars.

The star map also highlights several among the more visible stars that have name that NexStar controller utilizes as alignment stars. This lets you know whether the right stars have been selected or, if needed you are able to locate the correct ones yourself. The spelling may differ depending on the map or books, because the names for stars come from Arabic and aren't uniform.

The gray band is part of the Milky Way, the flat disc of the galaxy we are in. If we gaze at it we can see numerous stars that blur across the vast distance. This is a Milky Band which surrounds the Earth. In binoculars, we are able to disintegrate it into many stars. Nearby, we can also see open clusters of stars, as they are all located on the surface that is the center of our galaxy. In the star-deficient regions in the Milky Way, dark clouds of dust obscure the stars in their shadows. Dark clouds like these are the most visible when the sky is in summer, between Cygnus as well as Scorpius.

If we take a look from the space of our galaxy we'll see only a few stars, yet can see clearly the stars that are not part of that of the Milky Way: some globular star clusters (most of them are located within the direction of the galactic center, which lies located in the vicinity of that constellation Sagittarius within the sky of summer) and, above all, the distant galaxies. In the spring, we can see an enormous collection of galaxies within the constellations Virgo as well as Coma Berenice, however our galaxies in the vicinity are more to see within the constellation Andromeda in the fall sky, since they are further away from us. Andromeda is a galaxy that's close to us. Andromeda galaxy is easily visible with the naked eye when it is in favorable conditions. There aren't any planets on these maps. Since the planets are moving through the night, their location will depend on the date they were last seen. If you spot bright stars in the south, east or west, close to the red edge of the ecliptic which is not sparkling the sky, it's probably to be a planet. A yearbook, computer program, or an application like Celestron's SkyPortal will provide details on the planets that can be seen in what constellation and at what time. Additionally, you will find the time for setting and rising there.

Another alternative for these maps is planispheres. On them, you can choose the date and time using an ad-hoc rule and view the starry sky because of it. However the views of the sky are somewhat blurred: The north is depicted in a smaller space that the south.

In the star maps that are that are shown in this article, the center region is slightly smaller than the horizon region, but the orientation is clearer when viewed this way, even though a larger map has more objects on it and is easier to read because of the larger size.

These maps are only to give an overview. As when you need to locate objects with weaker features then you should look for an exact map, which illustrates, for instance, only one specific constellation. There is a variety of celestial maps. You can also make individual maps with your computer.

Looking at the Sun

Never look at the Sun without a secure solar filter!

If you put a secure solar filter on the front of your lens, you are able to have access to an amazing subject to study: The Sun, which fluctuates constantly and is also observable at all times of the day. It is easy to build the perfect filter for your telescope with AstroSolar film (see below for more details).).

With AstroSolar-film, you can make a solar filter by yourself and safely cover the lens.

The most affordable field of observations of the sun is called continuum, or white light. It is the portion of the spectrum we also can see with the naked eye. It is evenly reduced to a safe brightness with a highly effective filter that dims the spectrum of shades. The filter has to be placed on the close proximity to the lens, and block 99.999 per cent of the sunlight which allows you to see the sun's surface with confidence.

The sunspots can be the most prominent characteristic of the Sun however, they are not always visible. Around every 11 years there's an optimum in solar activity when there are many sunspots visible. Between these peak times, activity decreases dramatically and the Sun may be totally spotless for months.

In simple terms sunspots result from an increase in the magnetic field, which blocks convection. As a result hot gas bubbles not be able to rise to the surface from the deeper layers of the Sun and a portion of the Sun's surface is cooler than the surrounding. The only thing we can observe of the Sun is its red hot surface, which is almost 5,500 degrees Celsius hot surface. Much like a lightbulb, it shines because it's hot. The sunspots' temperature drops by 500 degrees, which is why they appear darker than the rest of the sun. They can be considerably bigger than Earth and, based on the size of their craters, are visible for a few days or even for a few weeks.

Additionally, they move in the course of the Sun across its surface.
The darkening of the limb is visible. The only portion of the Sun visible is the outermost layer of red heat (the "photosphere") which is from which visible light reaches. The light that we see from the edges of the Sun has to travel a longer distance over the photosphere as well as the layers that are above it (the chromosphere and"atmosphere" of the Sun) "atmosphere") which dims it more than light coming from the middle in the disc of solar radiation.
When the magnification increases the granulation is visible. Imagine the Sun as a large saucepan filled that is filled with water. Hot gases come from the inside of the Sun and transfer heat upwards. Cooler gas masses sink to the bottom. The gas bubbles are perceived as granules in the surface of Sun These particles can reach up to 1000 km long and extremely brief-lived. You require quiet, clean air to observe them in smaller telescopes.

The sunspots can be observed with a white-light solar filter.

SAFE SOLAR OBSERVATION

Using AstroSolar film from Baader Planetarium (www.astrosolar.com), you can easily create a safe lens solar filter (left picture). To create this filter, make a filter holder of cardboard , and then the safe film is adhered. It is placed over the lens in order that it's completely light-proof. If it is necessary, fix the lens with some tape so that it will not slide off. Remember to fix to the telescope that is used for viewfinder (attach the cap to the lens and secure it using the adhesive tape) or take it off. There are filters that come ready-made available.

Eyepiece solar filters, that were originally included with cheaper telescopes, are now in the garbage bin They are installed right in the focal point of the telescope. They absorb all the heat, which means they could (and could and) break and explode. If you gaze through the eyepiece when they burst, you'll have no chance of looking away quickly enough. If you use an optical telescope at the Sun without protection, permanent damage to your eyes is likely;

even looking longer into the Sun without a telescope may cause blindness. It is not recommended to try to cut costs by using a solar filter. One sheet of AstroSolar film is not anything, and it also protects from UV radiation that is invisible and can be quickly inserted inside an object-oriented filtering cell.

The projection of the Sun is a technique that is frequently recommended however, should be used with caution on the one hand there is the chance that someone could throw your warnings in the air and trying to see through the eyepiece. on the other side, the plastic parts or the black colour of the telescope may be damaged. Even if there is no damage evident, the smell of plastic could be stored inside the optics.

Alternately, it is possible to make use of Herschel wedges, however, only with refractor. It is recommended to select two" model, since small models can be more vulnerable to errors in tracking. For prominences (i.e. gas eruptions that occur near the edges of sun) you require a specific H-alpha filter, which is extremely expensive.

Never let your telescope go unattended while looking at the sun.

Variable Stars

There has been a lot written about the various types of objects that you can see with the amateur telescope. The book market provides many options of information However, you may reach a point that you do not want "only" to look your own eyes, but desire to explore more Astronomy, or perhaps contribute a little bit to the field of science.

According to me, two areas are especially suitable: Spectroscopy and the observation of stars with variable properties.

Surprisinglyenough, many stars change their brightness with time. Certain stars alter their brightness very slowly or over a long amount of time, whereas others alter in a noticeable way. Certain stars' curves of light are predictable while for others, they appear to be totally random.

SUCCESSFUL REVIEWING

For a successful observation of variably-colored stars, you don't require a huge telescope. In most cases, binoculars or the naked eye is sufficient. More important is having good star maps, and most importantly perseverance and patience.

In general the estimation of brightness is not a issue. You simply observe the star and contrast its brightness to that of nearby stars. It is important to choose only stars that have a constant and precise brightness as comparison stars as well as star charts that are backed by solid data on brightness. Simple atlases must be used with care, since they could only show an average of brightness moment the map is made but don't mark any variable stars as such.

One of the difficulties is to locate appropriate stars to compare. Particularly with bright stars, this can be a problem. Betelgeuze in Orion is one of the variable stars of long-period however, Betelgeuze is so that bright even at its smallest the brightness is so high that there aren't any other stars with comparable luminosity.

If you're looking to increase your technological investment, you could assess the brightness of the star using cameras. However, comparing it with observations from other astronomers isn't as simple, as cameras have a slightly different sensitivity from the eye , and has to be calibrated.

Light curves are the result of the variable stars.

When eclipsing binaries are used (above) The brightness decreases on the spectrum of light as stars are overshadowed by its counterpart. For long-period variables and Pulsation variables (below) the brightness increases depending on the size of the stars.

DIFFERENT STYLES OF STARS

There are a variety of reasons behind the fluctuations in brightness of stars. The variable stars are divided into the following types and, in addition, there are additional subtypes.

Eclipsing binaries can be easily predicted light curves because their brightness decreases with frequency each time the double star system moves ahead of , or in the direction behind other star. The most well-known example of this kind can be found in Algol (b Persei, the "Demon Star of Perseus"). The brightness of the star decreases between 2.1 to 3.4m each 2.87

days, for a total of 10 hours. Algol has been regarded as a variable for ages however it was only it was in 1783 that John Goodricke determined its period. Nowadays, the brightness minimum and maximums of a variety of binaries that eclipse can be predicted.

The ecliptic binary with the highest speed can be found in AM Canum Venaticorum, with the duration of only 18 minutes. Epsilon Aurigae has the longest known time period of 9883 days (27.05 years).

Variable stars that pulse are those whose size fluctuates regularly. Variable stars with long-periods (Mira star) are red, old giant stars, like Mira (o Ceti), which's brightness can vary by up to ten times in the span of between 80 and 1,000 days. The interval between two maxima are not always the same, and we can only make rough estimates of these star types.

The short-period cepheids are bright supergiants that pulsate. There is a constant relation between their duration and their brightness. The patron name in this group is called d Cephei. The brightness of this class varies between 5.4 days, ranging between 3.6 to 4.4m.

The semi-regular and irregular variables are red giants that pulsate or supergiants, whose light curve has no or little regularity. These stars are part of two distinct groups. On the other hand, they are long-period variable stars that exhibit extreme changes in brightness and duration between periods. The typical interval is only 200 or less days that span between two maximums. The other group of supergiants are young red supergiants, like Antares and Betelgeuze where nuclear fusion occurs throughout the hull.

The term"eruptive variation" is used to describe stars that exhibit unpredictability in brightness. This group includes novae as well as supernovae, as well as young stars as well as the R Coronae Borealis stars, where dark cloud of soot frequently block out views of the planet. They're a diverse group however, not any of the members are fully understood. Supernovae and Novae aren't identified.

CONTACT

The AAVSO gathers results from observations and is able to answer inquiries. You will also find information and maps as well in calls for observation. The more observers take part in a contest and the more they participate, the better the likelihood of getting creating a light curve that is as accurate as is possible. Further information can be found at: The American Association of Variable Star Observers, 25 Birch Street, Cambridge, MA 02138, USA, http://www.aavso.org.

Telescope Technique

What are the functions of a telescope and what kinds of mounts are available, and what should be considered when choosing accessories? What tools can help you maximize the performance of your telescope and what alternatives does the NexStar hand-control offer? Which telescopes are compatible with which accessories and which do you think are appropriate? This chapter will provide details on the different telescopes, as well as an idea of the various accessories available.

Chapter 11: The Refractor The Classical Telescope

When we hear the term telescope, the majority of us are thinking of the long refractor. Lens telescopes were initially utilized by Galileo Galilei in 1609 to observe the stars just a few years before it was discussed by Dutch manufacturer of spectacles Hans Lipperhey. In 1611, Johannes Kepler improved the concept somewhat, and today with modern varieties of glass, excellent telescopes are feasible.

The goal of a lens telescope is comprised of glass lenses that reflect light, thus the name refractor. The colour (or length) of light is refracted at a different direction, meaning that a lens works as a prism, and splits the light in all colors. Through the combination of two lenses, and making a smart choice of focal length you can create a clear image with all colors are focused at the same location with only a tiny amount of color fringe is left. Achromatic telescopes since they minimize the color errors.

Light path in a lens telescope

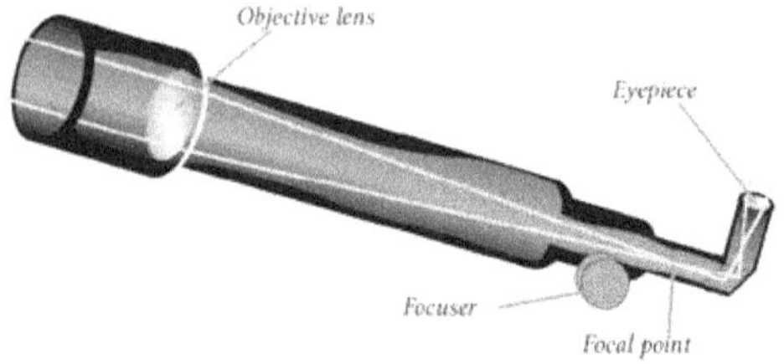

In the early 19th century Joseph von Fraunhofer figured out there was an extra distance that existed between two of the lens (the "air gap") and a smart selection in the aperture (lens size) along with focal length enabled him to create an image in which there is no color aberration with blue fringing around bright stars, or edges with high contrast disappears. The real, true-color Fraunhofer refractor has a very long focal length however, there are numerous achromatic refractors with an aperture that is large and a short focal length. These telescopes are commonly described as Fraunhofers due to the doublet lenses they have. They offer clear images and a vast field of view , so that finding the right spot to the sky is easy and, more importantly big objects like star clusters, gas nebulae and comets appear stunning. However, at higher magnifications for double stars and planets the color errors are typically visible as a blue fleck and the contrast is reduced.

Perfect Images

The most expensive apochromatic telescopes offer an image that is perfect, and displays pin-point stars with no coloring fringing even at high magnification. They employ lenses composed of costly glass or even fluorite crystals. the lenses typically comprise of three or four lenses.

The ED-Apos was introduced to the market in the middle of the millennium's turn The two objective lenses is constructed out of extra-low dispersion (ED glass). This makes excellent shorter focal length telescopes feasible with a perfect image where there are no colour imperfections evident. However, using ED glass by itself is not a guarantee of a perfect image. Every optical component needs to be in perfect order, as color and other mistakes could occur.

THE COATING

Alongside the lenses coatings, coatings affect image quality , and most importantly the brightness of the image. A range of four to five percent the light is reflective on the surfaces of an uncoated lens and can lead to a significant loss of light. Good coatings reduce light loss and scatter by less than one-percent.

The ideal situation is for a telescope to be fully multi-coated. That means that all glass-air surfaces are multi-coated. Multi-coated is the term used to mean there is at minimum one of the surfaces of the lens has been coated. Full coated implies that every lens must have at minimum one coating that is simple, and coated means that at most one lens has been coated on a single surface. The higher and more durable the coating is, the clearer, more brighter and more stunning the image will be when compared with an uncoated, identical telescope.

ADVANTAGES and DISADVANTAGES

Lens telescopes do not just appear like a layman would imagine a telescope to look however, they are also easy to take care of the lens is securely mounted in its cell and generally retains the factory adjustment. Only very top-quality refractors (and some less expensive ones) come with an adjustable mount for the lens, however, if the telescope has been meticulously designed and properly maintained an additional adjustment should not be needed.

They are, therefore, simple to maintain and produce extremely high-contrast images thanks to high-contrast stars with needle-sharp edges. Newtons and Schmidt-Cassegrains feature an additional mirror in the light path, which reduces the contrast. If you have an Apo or real Fraunhofer with a large aperture, the entire aperture can be utilized (there are no "obstruction") and the optics are able to be utilized to their fullest potential. Temperature adjustment isn't an issue at all.

The eyepiece is located at the back, which means you will require a tripod that is high and the diagonal of the star is adjustable to allow an ideal view from any angle.

The most significant drawback to this kind of telescope is the price as at minimum four lens surfaces must be precisely machined the cost is quite high and models with larger than 100mm aperture are scarce.

Image Origination

When using a lens telescope that does not have a star diagonal the image rotates 180 degrees, i.e. upside down and inverted. when a diagonal is a star the image is inverted but upright. When using an Amici prism it is vertical and laterally correct, just as we have come to expect with binoculars or an Spotting scope.

Chapter 12: The Newtonian - The Costly "Light Bucket"

A Newtonian reflector telescope is the highest price-performance ratio: the optics comprise only two mirrors. This means that the large light-collecting area can be achieved for a small amount of money. This is why they're sometimes called "light buckets." In the current form the reflector telescope dates all the way to Isaac Newton, who developed the telescope in 1668. There are, however, several other, somewhat unusual variants, such as Cassegrains (which look like the Schmidt-Cassegrain but do not have a Schmidt plate to correct images), Schiefspiegler (also called tilted-component telescopesthat have the same performance as lenses that are very high-quality, however they come with very large focus lengths) and various other types of constructionthat you'll see very rarely or not ever.

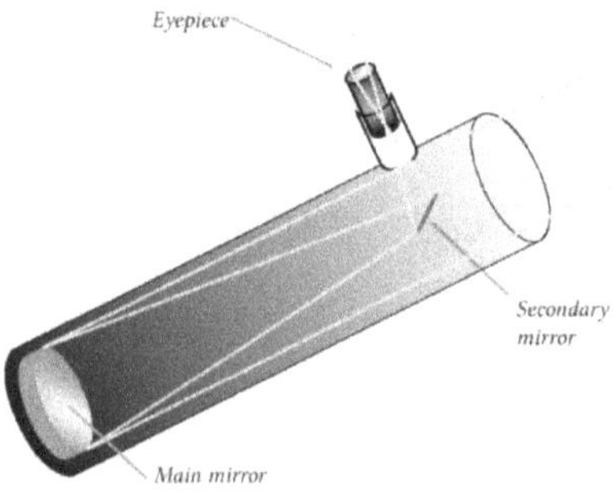

Light-paths in the form of a Newtonian.

It is recommended to be cautious when using catadioptric Newtonians The lens element inside the focuser expands to the length of its focus. This means they are small and compact, with a large focal length. However, the length of the tube in the real Newtonian is about equal to the focal length. This isn't in the end good, but it puts extremely high expectations on the mechanical components, which typically are not met by inexpensive equipment - and the cheap telescopes cannot be adjusted, because the exact mechanics are not present.

The most vibrant color FIDELITY

Because the light from the Newtonian reflection is formed by the light, and not refracted or split into colors like refractors and refractor, there isn't any hue aberration in the Newtonian. A parabolic mirror is required to gather all the light that is in the focal point. A simple spherical mirror can only provide excellent images when it has a tiny apertures or slow the aperture (such as F/8 or even f/10).

Particularly "fast" Newtonians with an aperture ratio of F/5, F/4 or faster display comas on the edges of the image. This means the stars appear to be inflated to the size of small comets. To get rid of this error it is necessary to use an coma corrector which is located directly in front of your camera or eyepiece. If you are primarily looking to see visually then using a Newtonian that has a f/6 or lower is a great choice since you don't need expensive eyepieces or coma correctors while f/5 is an option that was designed specifically for photography, however it is still able to work effectively with standard eyepieces, and lenses with f/4 and more are designed specifically for photography. The aperture ratio does not only affects the eyepieces usable, but as well the length of the tube. As the length of the tube increases it also increases the demands on the mount, as the lever as well as the susceptibility to wind are higher.

ADVANTAGES AND DISADVANTAGES

There are no color mistakes and great value for money are two reasons that the Newtonian is the most well-known instrument for introducing the field of astronomy for a long time. There is no other instrument that offers a greater than a light-gathering lens at the same cost and, on a basic Dobsonian mounting, huge telescopes can be made.

However, the Newtonian has some drawbacks as well. Secondary and main mirrors have to periodically be adjusted in order for the telescope to function at its maximum power. However, at least you have the ability to alter the adjustment in the factory. It is not always ideal even because the mirrors may move during transport. With a little practice and using tools like the laser collimation tool, or the Cheshire eyepiece or even a telescope that is large can be adjusted in just a couple of minutes.

The fixed location of the focuser is more difficult. You will always enjoy a pleasant view when using an alt-azimuth mount However, with an equatorial mount , the location of the focuser is moved across the sky thanks to the telescope. This means that the telescope needs to periodically be rotated within its clamps. This is why it makes sense to begin by observing objects in one part of the sky before observing objects in another in order to only need to rotate the telescope one time. A diagonal star as well as an eyepiece turret isn't feasible using the Newtonian as the focus cannot be achieved using it. In addition, using the DSLR camera, which has a the focus point is different than an eyepiece, could be a problem.

The"so-called "spikes" are four light rays that emanate from bright stars - could require some time becoming used. The "asterisk effect" is not a problem for image quality. It's caused by light diffracted onto the structure of the secondary bracket. If you notice photos that have the typical spikes, then you are able to determine the shape of your telescope. If you find the spikes annoying or it's something to be judged by your taste. certain astrophotographers make them appear more by stretching wires behind their refractors or by employing the correct Photoshop effect.

Newtonian telescopes generally have limited focal range. If they're made to be used for photography or visual observation, you'll need an extension sleeve that can be able to focus the telescope using an eyepiece. It's included with the telescope, however, it is not always installed.

Image Origination

The image of Newtonian Newtonian image is upside-down. One way of achieve an image that is upright is standing in front of the telescope to take a look - different optical designs are more suitable for observation on the earth.

Amici prisms are not compatible when using an Newtonian because they're too long and can't focus correctly. Simple erect-image eyepieces are available (at lower quality) but they're only available in sets with inexpensive entry-level telescopes.

The Schmidt-Cassegrain The Schmidt-Cassegrain The Versatile Telescope

Schmidt-Cassegrains are often claimed to be capable of doing everything at a time, and not do anything particularly well. There's some truth to this, however, I've come to see their versatility, and most importantly, as practical telescopes.

The Schmidt-Cassegrain dates back to German optician Bernhard Schmidt from Mittweida, who developed the Schmidt telescope or Schmidt camera in the year 1930. It consisted of a fast mirror with a glass plate as well as the camera, placed within the tube of the telescope. The 1940s, and the 1950s the system evolved further in the USA to allow for visual observation. A convex secondary mirror is able to reflect the light back toward the main mirror, and is released through an opening in the main mirror as well as the back inside the tube.

A light path in the Schmidt-Cassegrain

This catadioptric designthat incorporates mirrors and lenses, allows for extremely compact telescopes. Secondary and main mirrors are spherical mirrors that are fairly inexpensive which are in perfect alignment with each the other. They are not just matched in distance however, but also their alignment towards each other has to be maintained to give good pictures. If the plate was not correct they'd be useless telescopes that suffer from coma and wouldn't be able to provide sharp images. The corrector, also known as the Schmidt plate can be described as the more intricate and expensive part: for grinding, it's pushed into a mold in a vacuum and then processed. Only after the vacuum has been removed is it able to take on its final form. Even during times of mass production it is it is not possible to make the Schmidt plate and more so, the mirror that is secondary cannot switch between two telescopes. Furthermore, the correct orientation for installation should be maintained.

in the 60s Tom Johnson, President of Valor Electronics in California, was interested in telescopes. The goal was to create a powerful but compact instrument. After the huge public reaction to his initial 18" cassegrain, he began thinking about making a telescope suitable to make series production. He discovered a way to manufacture the Schmidt plates with less effort and also in bulk. In 1964, he established Celestron Pacific. It was Celestron an early company manufacture small Schmidt-Cassegrains with greater quantities and also to provide affordable powerful, high-quality and most importantly mobile telescopes for amateur astronomers.

ADVANTAGES and DISADVANTAGES

Because of its small size and large aperture, the Schmidt-Cassegrain is able to be operated with a compact and cost-effective mounting. It is a great choice for mounting. Schmidt-Cassegrain can be priced in the range of two types of mounts: a Newtonian as well as a refractor that has identical aperture.

You can also save money on eyepieces. Because of the standard aperture proportion of about f/10. Even the most expensive eyepieces perform very well and provide sharp images. This allows you to purchase the correct eyepieces based on your specifications.

Because of its focal length, small magnification and big exit pupil aren't likely to be. For those who live in cities, this isn't a major issue, since our pupils aren't always fully open because of light pollution in any case. To get the most wide field of vision, binoculars are more to use than nearly any other telescope.

The length of the focal length is an advantage when it comes to observing the planets due to the fact that the required high magnifications can easily be obtained without Barlow lenses or small focus length eyes. The image is less bright when compared to the case of a Newtonian or refractor having the same aperture because that mirror in the second is bigger however this is not a problem for photography.

Due to the tube being closed due to the closed tube, the telescope requires some time to adjust to the ambient temperature. It is therefore recommended not to begin with the highest magnification right now, but wait for the telescope cool off initially, then begin by focusing on targets which require less magnification. However the tube that is closed protects the mirror from dirt and dust.

The most notable feature of the new Schmidt-Cassegrain is the mirror's main focus. The main mirror moves through an adjustable screw located on the other bottom of the tube. The length of the tube is changed slightly and may cause the mirror to shift as the mirror sways in its holder. This is not noticeable visually, but it could interfere with photography. To ensure that optical elements are in their proper position and orientation, the focusing knob should always be turned counterclockwise while focusing. The mirror will move in opposition to gravity, and the shift is compensated.

Another unique characteristic is the eyepiece connection instead of a conventional focuser, Schmidt-Cassegrains feature an eyepiece thread (either one of the "SC thread" with 2" or a bigger 3.3" thread) to which large equipment, like cameras that are large are able to be screwed. When using the correct adapters, the optimal space between mirrors is ensured, which allows for the highest quality image.

To attach eyepieces, you can use eyepiece adapters such as the "Visual Back" eyepiece adapter however it is only available at 1.25" using the Schmidt-Cassegrains. Baader Planetarium also offers high-quality 2" Clicklock eyepiece clamps, particularly on alt-azimuth-mounted telescopes the 2" diagonal of stars with an identical thread can be attached directly onto the telescope. Photographers can also use an extension to get the best distances.

Image Origination

Similar to the refractor the image rotates by 180deg with no star diagonal or laterally reversed by the star diagonal. Also, Amici prisms can be employed to ensure that the telescope is also used to observe terrestrial objects. Because the magnification is high, it only makes more sense with smaller telescopes. the C5 with 127mm aperture is also available with a 25mm eyepiece, which is an spotting scope, with 50x magnification.

Chapter 13: The Edgehd Clear To The Edge

From the 60s onward, very little has changed in the look of the Schmidt-Cassegrain however, the quality of coatings and production have improved , while the prices have decreased. To meet the increasing requirements of modern digital cameras, with their large and high-resolution sensors However, it was essential to modernize the design.

Celestron could have chosen to go with a completely new design or adapt the Schmidt-Cassegrain model that has proven successful to meet new specifications. The design team chose not to go with a totally new design, partly for reasons of financial - after all, the telescope must be affordable. In various studies and prototypes, models from Dall Kirkham Cassegrains have been made. They were priced too high for mass-market use, however, the research led to the establishment of the business Planewave which makes excellent "astrographs" specifically designed for photography, but at the same price range.

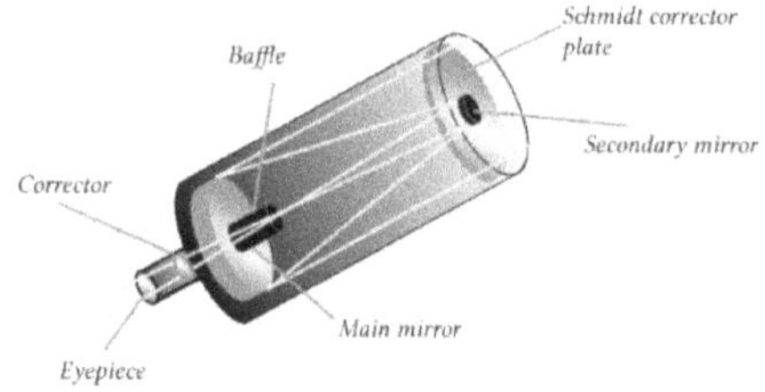

Light pathway in an EdgeHD

The latest Celestron telescopes are designed to provide an unbeatably smooth and evenly lit image field, which means that even the largest CCD sensors and Full-frame DSLR cameras can be utilized without image distortion that occur at the edges similar to the contemporary ultra-wide angle eyepieces, which have greater than 100deg fields of view. The objective was to develop optics that are diffraction-limited across the entire image area, i.e. that the quality of images is affected only by the inevitable diffracted light beams that are at the edges of the lens, whereas the other image-related errors are not as significant. This was made possible by EdgeHD. It achieves this with EdgeHD telescopes that permanently install two-lens correctors close to the optical back. This removes image distortion and gives more than 42 millimeters of image diagonal, without irritating comas, color errors as well as image field curvature. If the curvature is not corrected of the image field the stars near the edges of the image would become blurred.

To ensure that the optics operate according to plan and to ensure that the optics work as planned, the tube was modified. Apart from that it was the bearing on the Schmidt plate (which can now be adjusted with grub screws , and then glued to its final location) as well as the mount for the mirror's main mount were modified. Like a typical Schmidt-Cassegrain Mirror, the main one of the EdgeHD is moved to focus and guides rails on the EdgeHD have been greatly upgraded so that mirror movement is no longer a problem. In the traditional Schmidt-Cassegrain mirror, the mirror may move slightly, and this is not noticeable visually, however it can be a nuisance photographically. With EdgeHD, however EdgeHD it is tilted at smaller than 30 arc second, therefore it doesn't have any effect on the image. The mirror can mounted using the main clamp for mirrors.

Mirror shifting can be prevented with the use of a fixed mirror for the main mirror, but an additional, reliable and stable focuser would have significantly added to the overall cost.

The EdgeHD has air vents, which are shielded from dust with a metal grille that allows the telescope to adjust its temperature to the surrounding temperature, which reduces the tube's vision.

VERSATILE OPTICS

All EdgeHD are compatible with using the HyperStar system. That means the mirror in the secondary is easily replaced with an appropriate corrector system. The telescope also has the aperture of 2 f/2 which means that extremely large image fields and very small exposure times are achievable. The conversion process is quick so that the final product is an electronic Schmidt camera that is identical to the system that was originally designed by Bernhard Schmidt, from which the modern Schmidt-Cassegrain came into being.

If you're not looking to shoot in a huge field of photography, the 0.7x reducers provide a excellent opportunity to reduce the time of exposure in comparison to the entire focal length, and also get a bit more space for viewing. Reducers don't cause additional vignetting using full-frame cameras. They use similar adapters to photography that does not have Reducers The distances between your Reducer and the camera's lens are exactly those of your telescope and camera. The only difference is that the 8"-model requires a shorter distance, however the EdgeHD-T-adapter can be divided in half to provide the correct distance. Barlow lenses allow you to get the maximum magnification needed for planetary photography.

ADVANTAGES AND DISADVANTAGES

One benefit of mirror focusing technique is that the optics perform optimally when a certain distance is between components. A second focuser should therefore be used only within a certain range and should be configured for photography so that the camera is placed at the optimal spacing that is 5.75" (146.05 millimetres) as well as 5.25"

(133.35 millimetres) to accommodate eight" EdgeHD. In the absence of an extra focuser, secure connection and the proper measurement for SLR cameras is guaranteed by EdgeHD-T adapters made by Celestron, because the distance between contact points from T adapters is established. For extreme CCD cameras, appropriate adapters are required; the distances are kept to +-0.5 millimetres. However, smaller tolerances are feasible, depending on the size of the chip.

If you are planning to utilize a focuser, it is recommended that you place the telescope in the direction of the star to ensure that a camera located at the proper distance will give you sharp images and finally, make sure to lock the mirror. That is the way the telescope operates in its optimal arrangement.

If you are only observing visually and don't use ultra-wide angle eyepieces then you can opt for the classic Schmidt-Cassegrain with no hesitation. Wide angle eyepieces are generally not a problem for Schmidt-Cassegrain, as you typically focus on the middle of the image, not towards the edges of the image. EdgeHD EdgeHD showcases its benefits for photographers. Sharp-edged images as well as the existing Reducer and the process for HyperStar make it an intriguing system.

Alt-Azimuth Mounts that do not require Computer Control

Alt-azimuth mounts that do not have computer control are great for those who want to learn about the sky yourself and also if you want the convenience of a telescope that can be dismantled and set up quickly. This is particularly useful for those who want to go back to the warmth after a winter's cold night.

The term "alt-azimuth" refers to the kind of tripod: it is adjustable in the direction of altitude (up/down) or azimuth (right/left) similar to the two-way panhead of the tripod for photography. It is operated with ease and all you have to do is set the tripod or the mount on an even surface.

There are three major kinds of azimuthal mounts.

PowerSeeker 80AZ-S is mounted is mounted on an Alt-Azimuth Mount with the ability to finely adjust height.

1.) For alt-azimuth mounts that have the ability to adjust finely (picture above) the telescope is inserted inside a fork. Typically, an aluminum rod with an incredibly small wheel to adjust the fineness of the to the mount and telescope. To find objects, this rod is let loose so that you can rotate the telescope in a swiveling motion; after you have located your target, secure the rod or axle. The knob that is ribbed on the axle permits users to make a precise adjustment to ensure that your targets remain exactly within the eyepiece when you are using a the highest magnification. This adjustment is only for the height axis. You may choose to let the azimuth axis clamped or tighten the clamping screw to the point that you are able to move your telescope using only a tiny amount of force. It is desirable for the telescope to move without resistance, but it remains in its location after you release it. This makes it simple to align objects in nature observation. In astronomy, it is easy to compensate for the movement of the stars.

AstroMaster 70AZ mounted on a simple Alt-Azimuth Mount (two-way panhead) with a long adjuster handle.

2.) Telescopes equipped with a two-way panahead (picture above) operate like cameras on a quality tripod for photography. The long lever such as utilized on AstroMaster tripods can allow you to follow the telescope with precision. Because the telescope is a tripod, you will not require a table for an base. You you can place the tripod on the ground. Particularly if you own an optical telescope, you must bring your chair along with you. They are extremely large, and therefore the eyepiece is low. And when you are sitting at a table, you can observe more comfortably.

FirstScope 76 on an alt-azimuth table mount - a tiny Dobsonian telescope.

3.) Three. The Dobsonian mount (picture above) was designed around 1950 in the 1950s by John Dobson as a simple and affordable mount. Even the largest telescopes are easily mounted by using a huge box inside which the telescope that has the altitude axis is placed; the box (the "rocker box") can be turned in conjunction to the scope. This mount is particularly well-liked by Newtonian telescopes since the eyepiece isn't rotating - it is always viewed through the eyepiece through the top without needing to rotate the tube. The mount is utilized as an example with the FirstScope 76 The only thing you have to do is put it on a sturdy table and then observe. The axes are clamped in any way (in the azimuth) or are only clamped so lightly that there is a sliding clutch and the telescope rotates without jerking whenever you push it on , and it remains in its location.

Overview: AL-AZIMUTH MOUNTS

Advantages:

* Quickly ready to use
* Lightweight

Disadvantages:

* No tracking system is automatic.
* No deep sky astrophotography using lengthy exposure time possible

Equatorial Mounts with no Computer Control

The the equatorial or parallactic also known as German mount, also known as the German most common mount used for astronomy. A single is an axis (the right ascension axisor R.A. axis or polar Axis) is aligned with the Earth's axis. Thus, it is pointing in the direction of north to the North Star (more precisely to the northern celestial pole where all stars revolve). Contrary to an alt-azimuth mounting, only this axis needs to be moved in order to keep an object inside the lens. If you do return to the telescope after a time you'll be able to find the object once more Simply move the control of slow motion for the R.A. axis in the proper direction, and it will appear within the lens.

PowerSeeker the 114EQ is mounted on an equatorial mount. Set at lower elevations.
In order for this to function it is necessary to be able to align your right ascension axis as precisely as you can to the north, and then adjust the altitude to your specific latitude. It is important to establish it: verify the

location of north (either by using a compass or by searching for Polaris or a simple compass can guide you to the magnetic north polebut not precisely to the geographical north pole) and then set the mount with north. The altitude setting needs to be set to reflect the latitude that you're observing (e.g. e.g. around 50deg in Central Europe. So long as you don't try atropotography, you will not need to align exactly however the more you're off the more frequently you'll have to correct on both axes, if you are watching for long periods of time.

To ensure precise positioning for precise positioning, there's the telescope called a polarscope (#94223) for Omni telescopes. It's mounted on the mount first, you must set dates and times and then move the telescope to precisely capture constellations, such as using the saddled camera. To observe with a visual view, put your telescope at its standard location (as illustrated in the photos in this article) and put the Polaris at the center of the eyepiece or finderscope.

It is also crucial to balance the telescope first let go of the clamps on the right ascension axis , and move the counterweights till the telescope stops moving itself. Then clamp them up again and then remove the declination axis and then move the telescope into its pipe clamps till it's balanced.

Omni XLT 120 on an Equatorial mount, designed for mid-latitudes.
Since the axis is tilted about 40-50 degrees, shifting the telescope and locating objects can take some time to get used to. Once the clamps are loose you can become familiar with the way the telescope operates. It is also possible to rotate it in case you are following an object that is high in the sky, from east to west, or the telescope might be struck by the tripod or mount.
The benefits of equatorial mounts are beneficial if
* be sure to observe at high magnification or for a prolonged period,
* observe in groups, and allow other people to observe through the telescope or even if you

* would like to shoot astrophotography using longer exposure time.
To benefit from an equatorial mount, buy a tracking motor (#93514 for easy tracking of the AstroMaster and PowerSeeker mounts #93522 to track each of the Omni Mount axes). But, you should disengage the motor whenever you rotate the telescope using the mount's slow-motion control; for larger swivels, just remove the clamps on the axle just like normal. With the motorisation on two axes of the Omni telescopes both axes can be adjusted to a specific degree using the electronics using the handheld control box.

Simple tracking motor that can be retrofitted.
If you are able to achieve an altitude of zero degrees typically, you can make use of it as an Alt-Azimuth one as well. This is great for nature observations or for a quick glimpse at the sky. It will increase stability, so that the image appears more peaceful at higher magnifications.

Conclusion

The equipment used in astronomy is costly, therefore I hope this guide has provided you with the necessary information to make the right selection of the equipment that best suits your requirements.

Lightning Source UK Ltd.
Milton Keynes UK
UKHW052138251122
412725UK00021B/1653

9 781774 854495